EVERYTHING THAT RISES

Love is climate action!
Vote. Protest. Divest.

[signature]

EVERYTHING THAT RISES

A CLIMATE CHANGE MEMOIR

BRIANNA CRAFT

Lawrence Hill Books

Chicago

Published by Lawrence Hill Books
An imprint of Chicago Review Press Incorporated
814 Franklin Street
Chicago, Illinois 60610
ISBN 978-1-64160-860-2

Library of Congress Control Number: 2022948633

Typesetting: Nord Compo
Map design: Chris Erichsen

Printed in the United States of America
5 4 3 2 1

CONTENTS

PROLOGUE
THE DREAM

Kelso, Washington
1990s

MY STRONGEST EMOTIONAL MEMORY of the man who lived in my parents' house was fear. The stuff of nightmares. One, in particular, always began the same way. There was some kind of ruckus: people shouting, fighting in the dark. I didn't know where in the house I was yet.

Then I would hear myself scream.

The noise would pull the scene together. I was running. Down the slick wood of the hallway, through the door into the garage. Nighttime, so the only light came from the heater by Mom's car. Usually wet with condensation, the metal doorknob to the back stuck and was hard to open. It squealed when I turned it free. Beyond, the woodshed's logs were stacked high for winter, but I wouldn't make it up to the roof in time.

He was gaining.

I ran out into the black of the backyard. Away from the Gardners' place—the only other house visible from ours, the two highest on the hill. The light on their porch turned on when animals crossed it, and I needed the dark. The grass was wet, but the clouds were missing. No white mist tonight. No cover. He would see. I needed the trees,

1

the thickest of the woods. Breathing hurt before I reached the deer trail through the blackberry bushes that, taller than me, bordered the lawn. I didn't know why I thought I would be safe there.

I wasn't moving fast enough.

Dad was chasing me.

He had a knife. And I knew I was going to die.

———————

The terror choked me awake. My outstretched arms searched for Big Bunny, the stuffed rabbit Granddad gave me. Big Bunny wasn't bigger than me anymore. I pulled him in easily and took big breaths into his fur until I could count the ticks of my alarm clock. When my eyes had adjusted to the dark, I traced the familiar shapes of the room I'd always had—the tiger poster above my bed, the books and toys, the soccer ball. My *Little Mermaid* shirt on the floor.

Everything was quiet.

Nothing from my only sister's room next to mine. Chanteal was two years younger and she slept better than me. Nothing from farther away, upstairs where Mom and Dad slept. I couldn't hear the road, even when the gravel flew on busy days. The only noises at night were train whistles from town, and I usually had to be on the roof to hear those. Up there the view was east over the hills of trees to Mount Saint Helens, who blew her top.

Outside my window, the firs moved gently in the wind. Clouds covered the sky, unbroken. It rained every day this time of year. Not yet today though.

I probably wouldn't be able to sleep again. Sometimes, I called for Mom and she crawled into bed with me. But not when the dreams were about Dad. Calling might wake him up. And Mom and I never talked about why I had nightmares. She didn't want to hear it. Afraid, I buried my face into Big Bunny again, trying to hide. He needed a bath, but I breathed into his fur anyway.

Dad didn't like me. Not like Chanteal. He loved her. He didn't hit her like he hit me. He didn't hit anyone like he hit me. He left no one else crying on the floor. I was always making Dad angry, and at seven,

I didn't understand why. Tensions grew until there was a confrontation. After, I would have bruises. Or Dad would stop talking to me. Sometimes both.

Most people didn't know this.

When he wanted to be, Dad could be really nice. He wore a suit and worked at the hospital. He had friends from the YMCA, where he played basketball and racquetball. He shook hands with everyone in church. The neighborhood parents who played golf together said he was their favorite whenever he told jokes or laughed, showing all his teeth. "Your dad's a hoot," they said.

I never told them what he's like, to me.

I mostly didn't talk about him, or I told stories that made me seem loved. I pretended. And sometimes, I lied. I really wanted things to be different, and I really didn't want how things were to be true.

The glow-in-the-dark arms of my alarm clock said it was still hours until wake-up time. Then Dad would take Chanteal and me to Longview Christian School. We'd be there for breakfast. School breakfast was fine, but the lunches were gross, except for on Tater Tot casserole day. That's why Mom made me peanut butter and jelly sandwiches to eat instead. They were good all mushed down thin.

Inspector Gadget would come on soon, then *DuckTales*. I couldn't get up to watch them, even though the antennas were already in the right place. I wasn't allowed TV before the alarm went off. Instead I slowly uncurled to look out the window again, searching for stars.

I thought them magic.

Stars were my favorite things about the world. Probably because I didn't get to see them a lot. When my guinea pig Whiteface died, I imagined she could look down on me through the stars. It was a made-up story, like Santa Claus coming at Christmas. And the clouds were too thick to find her tonight. Still, I looked anyway. It would be so awesome to be up there, protected, to look through windows back to Earth, to get to see everything for what it was.

I loved thinking about that, loved the freedom of the imagining. Sometimes, it even sent me back to sleep.

1

A BEWILDERING INCEPTION

Providence, Rhode Island
September 2011

I CAME FOR THE FREE SANDWICHES.

A month of grad school in the Ivy League had left a grand total of $120 in my bank account. I was determined that no more would leave it until my next student loan deposit, which promised a somewhat less terrifying balance. So when my advisor mentioned a weekly discussion group and that lunch would be provided, I was there, early.

I waited in the sun-soaked room of a gardened brick building, watching the dust swirl in through the windows of a Wednesday afternoon. Autumn pulsed with the diminishing heat of the Rhode Island summer, my first in the Northeast and about as far from the cool breezes of the Northwest as I could get without actually leaving the country. I liked the way the heat warmed everything here even through to the edge of October.

I smiled at Becca when she came in. Becca was my first friend at Brown University. She seemed normal, which, in the context of the five women who made up my cohort of first-year grad students, meant she wasn't from money either. I understood that. We got along great.

Our convenor was next through the door. Professor Roberts set a bag of buns down on the large round wooden table occupying the center of the room and moved away from the hands greedy to disassemble its contents. Then he warmed up the projector and got a second-year up to present his PowerPoint. We were meant to talk through the converging challenges of climate change and development, not just whose turn it was for the tuna melt, though this definitely occupied my attention for the first several minutes of our hour together.

"Warmer temperatures and increased ocean acidity are clearly linked to decreasing oyster takes throughout New England," the second-year was saying when I stopped crunching long enough to listen.

This held my attention. These were the real-world impacts of the climate crisis, the things I had crossed the country to learn how to address. After an AmeriCorps year spent teaching kids in after-school environmental clubs, removing invasive species, and organizing local climate co-ops in Seattle's south side, I knew there was no going back, no more ignoring what I wanted. Doing something about the global threat affecting everything and everyone was more important than the regimented future continuing my architectural aspirations afforded. At twenty-four, I would go back to school. I decided to gamble two years and all the borrowed money I could muster on a master's in environmental studies in the fabled Ivy League, where I hoped to turn passion into a bankable career, one that would make a genuine difference. This was a tall order for someone whose experience of private education had ended with fifth grade. But school was school, and I was good at it. And I was determined to make the most of this.

From my mismatched chair, I took in the last slide of the second-year's presentation as it grinned down at us. It was a photo of him in hip waders holding a rusty oyster cage, surrounded by the growers he'd spent the last two years interviewing. He shuffled the pages in front of him in closing, and I joined in the round of appreciation and applause.

Professor Roberts leaned forward to comment before the clapping stopped. "That's fascinating," he enthused, interrupting the presenter's thanks. "Did your data collection reveal the same trend in farmed clam populations?"

I looked across the table at Becca. I thought my lanky, spectacled advisor epitomized the academic stereotype. He was so unlike the smooth-talking crew of architecture professors I had known before, who sent us to ask contractors about building materials and people about public spaces. I still didn't know what to make of Professor Roberts. Did people really never leave the Ivy League? Becca caught my eye, smirking. Maybe she thought the same. After completely missing the answer, I went back to taking notes.

The last item on the agenda, "any other business," materialized as the potato chips circled the table for the last time.

"Let's see. Oh yes, there is one note here," I heard over the noise of people closing notebooks and pushing back chairs to leave. I had a hand on my pack, which had found its way under the table. "I've been contacted by a London-based researcher at the International Institute for Environment and Development."

I felt my eyes bug as I straightened up. My advisor was apparently in touch with a slew of international researchers working on exactly what I hoped to. This place was wild.

"She'll be advising the chair of the Least Developed Countries Group at the next UN climate change negotiations, and she needs an assistant."

The United Nations climate negotiations? Where governments made legally binding decisions to address the crisis? Stuff of research papers and biannual mentions on *PBS NewsHour*? This wasn't the chance to work with a Seattle firm for the summer or maybe swing a shadowing trip to California. Public school didn't have this either.

"Any takers?"

I was half out of my seat, hand in the air. Just one among the limbs of suddenly attentive overachievers, mine was first.

"Oh, well, it's good to see so much interest," Professor Roberts said, looking around the room. Perhaps opportunities like this weren't exactly ordinary. "I think Brianna had it."

I grinned wide and straightened to standing, unable to believe this entire sequence of events. The second-year sitting across the table from me cursed, which brought me somewhat back to reality. I laughed.

"I'll put you in touch with her, Brianna." He typed the email as he spoke, and I walked round the table to catch his words over the resumed noise of people filing out. "You'll need to travel to South Africa a week early, but you'll see so much more as a member of a national delegation."

"No problem at all," I beamed, faking understanding. Then, because I simply couldn't keep it in, "This is so amazing! Thank you, Professor."

"I'll leave it to her to give you further details." He typed away, seemingly unfazed.

I was smiling like an idiot, absolutely elated. This was why Brown was going to be worth it.

I was going to the United Nations!

———————

Several weeks later, I landed in South Africa with remarkably little information. I knew I was headed to Durban, a city on South Africa's east coast. I knew the researcher I sought was named Achala. And I knew that we would start work on Monday. That was about it. Even so, excitement prevailed. I loved an adventure, particularly in places where blending in was a real possibility, and Africa seemed an excellent destination for anonymity.

I quickly learned that South Africa was a long way away. The drowsiness induced by the fifteen-hour flight from New York to Johannesburg and the connections on either end completely overpowered me. I remembered getting into a taxi outside the airport. I did not remember the journey to the beachfront hotel where I woke hours later, finally having reached my destination.

Groggy, I pulled back the floor-to-ceiling curtains. The cars below drove on the left side of wide, paved streets, running between buildings of diminishing height as they extended inland. A low rise of green hills framed the distance. The angle of the sun told me it was late Sunday afternoon, which meant I had hours to explore this city new to me.

Work first, though.

I flipped open my computer and tried the Wi-Fi. Sluggish success, but my inbox was disappointingly quiet. The email chain Professor Roberts had started, introducing Achala and me, left off several weeks ago— at least from her side—so I added a chatty announcement of my arrival.

Hi Achala,
Just wanted to let you know that I've made it to the hotel in Durban. I'm in room 1903 and would love to meet whenever you're free.
Looking forward to seeing you!
Brianna

I expected a speedy reply. Surely she would want to connect and go over what we were supposed to do so that we could hit the ground running tomorrow morning.

I waited. Watched some TV. Watched the sun start to set.

Still nothing.

When my stomach started growling, I gave up and headed for the lobby. After several unanswered rings to Achala's room, the receptionist shook his head.

"Sorry, miss."

Well, I was officially out of ideas. The view beyond the lobby looked amazing, and the receptionist's good eats recommendations sounded too nice to put off trying any longer. My mind made up, I finally headed out. The hotel was separated from the sand by a road, a row of palm trees, and an undulating pedestrian boulevard. I smiled at the heat on my skin, the December summer sun replacing concern with the delight of exploring. The beachfront was a lively place, full of families enjoying the coastline and vendors selling everything from beachwear to ceramics.

I spent the evening wandering up and down the shore, looking for sailboats beyond the long docks that jutted out into the waves. The life of a traveler was great. I had missed the freedom of knowing absolutely no one and having no agenda except to see a new view, experience a new culture, and eat something delicious. In my enthusiasm, I lost track of time and missed the phone man at China Mall, who vendors told me was the nearest local SIM card merchant.

When I made it back to the hotel, I tried calling Achala's room again. It was early evening, but jet lag wasn't going to let me stay awake much longer. I listened to the phone ring out with disappointment. The prospect that I had traveled across the world to hang out by myself seemed increasingly likely. I had no idea when we were due to start work tomorrow or where I needed to go. And as fun as it was, I wasn't here to travel; I was here to make the most of this ridiculous opportunity, which I couldn't do if I missed my first day.

Thinking my best strategy was to catch her early, Monday morning at 6:30 AM, I knocked on a stranger's hotel room door. I discovered that meeting your new boss in her pajamas, hair disheveled and looking at you like she wished you would disappear, wasn't the best way to make a good impression.

Whoops.

She was more gracious about it than I would have been. After realizing who I was, she let me in and cleared a chair of clothes so that we could chat while she climbed back into bed.

We met properly over breakfast an hour later. Once dressed, Achala presented herself as an immaculate native of Sri Lanka who had worked in the UN climate change negotiations for several years. A lawyer by trade, she had a petite, barely five-foot figure that made my five-foot-seven frame seem a stretch. She wore her dark hair shoulder length, burgundy lipstick, and heels whose tapping clicked rhythmically against the drone of her rolling bag. In the sea-facing breakfast room, I awkwardly joined her table. I hoped that the black Banana Republic dress bought on clearance the day before my flight was business-appropriate enough.

"Good morning again," I said, smiling sheepishly. The anxiety that drove me to wake a stranger seemed rather silly now. It vanished the instant we connected, leaving only embarrassment. "I'm sorry about earlier."

"That's fine," she responded, without making eye contact.

Of course I wouldn't have missed my first day of work. She would have contacted me in her own time. I sat waiting, watching her butter toast. She chewed, and I looked out at the whitecaps in the distance.

Achala's question broke the silence. "Where are you from?"

"A small town in Washington State. It's near Seattle, on the West Coast." I listed landmarks until she signaled recognition.

"And this is your first negotiation?"

I nodded, chasing a pineapple clump through a bowl of yogurt with my spoon. I was not a fast eater and, especially when nervous, I tended to move things around rather than actually consume them. I left food uneaten more often than I liked. My parents tried to school this trait out of me, but their attempts usually ended with me spending hours at our deserted dinner table, memorizing the contours of long-cold Brussels sprouts.

"I'm really excited to be at the negotiations," I chirped, eager to impress. "I've wanted to see them since I started learning about climate change."

I assumed Achala felt the same. Perhaps this would be the common ground between us, the space where talking wasn't so much work.

"I started in the UNFCCC negotiations as a student too," she said. "This is my first COP as legal advisor to the chair of the LDC Group."

While she spoke, I worked out the acronyms in my head. "LDC" referred to the forty-eight Least Developed Countries. They negotiated together as a group. The "UNFCCC" stood for the United Nations Framework Convention on Climate Change and the "COP" was its annual Conference of the Parties, the fun title member states gained after signing a treaty.

"It's a tremendous opportunity to work directly with the negotiators. I'll need your help keeping the chair's schedule, taking notes, editing documents. What are you studying at Brown?"

"Technology transfer," I said.

I expected approval. A lot of thought went into what, specifically, I wanted to study. Definitely not how bad the climate crisis was—I didn't need more nightmare-inducing data for which I didn't have the scientific education to interpret. In my pre-Brown reading, I gravitated toward the tangible means of combating climate change. Getting countries to increase the efficiency of their energy systems while transferring to solar panels, wind turbines, and other forms of renewable energy seemed the most inspiring.

"Oh," Achala sighed, "the worst one. Why do you study that?" She rose to leave.

Her disappointment threw me, even though she hadn't asked the question harshly. I stammered out a "Well, um . . . when I think about climate change, environmentally sound technology is what I think about."

Blinking, I followed her out the hotel doors into the sun. I knew I hadn't made my point. "It's interesting to me," I said.

Achala looked doubtful and said nothing more on the subject as we climbed into a taxi. She told the driver to take us to the international convention center, then told me that we would need to register to access the venue. The LDC Group met to prepare before the negotiations began, which was why we were in Durban a week before their start date. Like her, I would be registered as part of the LDC chair's national delegation, The Gambia.

A quick Google search had told me that The Gambia was a small country on West Africa's Atlantic coast bordered entirely by Senegal, population: two million people, none of whom I'd had the pleasure of meeting.

We found registration outside the convention center in a white tent, whose floor nor ceiling was substantial enough to keep out the midday sun's effect on an old parking lot. I watched Achala present her passport. The man behind the registration desk typed and nodded. The printer beside him churned out a small badge of laminated paper. He attached this to a UNFCCC-branded lanyard and passed it to her with a "Thank you."

I came forward to do the same.

"Which country?"

"The Gambia." I tried to sound convincing. He scrolled through his computer monitor with a blank expression. Sweat trickled down our foreheads while I explained my last-minute addition to the delegation, and he stated that my paperwork hadn't gone through. We faced each other over the impasse. Achala added a more credible retelling, but she, like me, was an outsider to the Gambian delegation. Things were going nowhere fast when Bubu rounded the corner.

Do you ever know instantly about people? Almost like you've met them somewhere before? For me, so it was with Bubu. He entered, jacket-clad, waving at Achala. Aged somewhere between my father and grandfather, Bubu appeared to know everyone—in this case because his involvement with the climate negotiations stretched longer than I had lived.

The registration official was no exception. "Ah, Bubu, my brother!"

Bubu's word that my papers would arrive was all it took. I smiled for the camera and seconds later a pink PARTY badge bearing a tiny, terrible picture and the word GAMBIA under my name hung around my neck. Registered, I crossed through a ring of security fencing. Then I followed Achala and Bubu, who seemed intent on acquainting themselves with the jumble of temporary and permanent structures that would form our workplace for the next few weeks.

"So, you are from The Gambia now," Bubu stated. His smile stretched his lips over broad teeth. One of his central incisors had a sawlike chip.

I laughed, grinning all the while. "I suppose I am." Bubu's joviality was catching.

We fell in step behind Achala, who clicked purposefully forward looking for tomorrow and Wednesday's meeting room. The convention center was huge, full of echoing concrete corridors and gaping expanses that surrounded curved rampways up to other levels. I half expected to find an NBA game kicking off. The place was larger than I thought necessary to talk about climate change.

I changed the tenor of my voice to sincerity and thanked Bubu for helping with my registration. Achala joined in, then began pointing out things I should familiarize myself with.

"Brianna, these are the screens that will show a live feed of where the meetings are," she said and waved at a pair of now-blank TV monitors mounted to a wall.

I nodded, feigning familiarity.

"There's the documents counter where they'll have the latest versions of decision text."

It took us half an hour to locate the rooms we needed. The room for the LDC Group's preparatory meeting was large and full of people setting up tables. The LDC office was small and full of people setting up desktop computers. I tried to sear the locations of both into my mind, but it was the ugliness of the office's orange carpet that really stuck with me. That and how empty the colossal halls of the convention center were.

The next morning, I followed Bubu and Achala into the preparatory meeting with a shocked expression. So many faces waited for us.

The forty-eight countries* that formed the LDC Group were classified by the UN as the world's poorest. While passing row after row, I tried to recall the stats—attempting to name each country beyond me. Thirty-four were in Africa: spanning the Sahel, from the western coast to the eastern horn, and down through the Congo basin. Nine extended

---

* There's a map and list of the Least Developed Countries in the appendix.

from the deltas to the mountains of Southeast Asia, right up to the roof of the world in the Himalayas. Four were low-lying island nations spread among the vastness of the Pacific Ocean. And Haiti was the lone country that represented the Caribbean. A handful of negotiators from each of these countries made up their nation's party to the talks.

We claimed seats at the front row of tables aligned classroom-style toward an elevated platform. Tabletop microphones, spaced evenly between every few seats, faced the gathering of suited delegates. In front of them, most had placed a laminated fold of white paper that in clear, black capitals spelled out the name of their country. The man seated at the top table behind the name plaque that read THE GAMBIA glided down to greet us. Tall and angular, Pa Ousman smiled over the top of his glasses as he shook hands with Achala and me.

This was the LDC chair.

"Good to see you, Achala," Pa Ousman said. He continued the round of greetings with an enthusiastic clap of Bubu's hand.

"Did you get the notes I sent you?" Achala asked.

The transition from pleasantries to business happened too quickly for me to join the conversation.

"Yes. The ambassador is coming at two o'clock to brief us," Pa Ousman began. "South Africa's job won't be easy." I smiled politely, unsure if they wanted me listening. "We need a decision on long-term cooperative action."

"We also need a coordinator to lead those discussions for the group," Achala interjected.

Pa Ousman stopped, looked at the full room and then his watch. "We should start." I watched him and Bubu climb the stairs to the top table.

I felt out of depth and place. Yes, I was studying climate change. But my utter inexperience with the UN made me self-conscious, which unhelpfully duplicated my existing sense of self-awareness. The overwhelming majority of the hundred or so people in the room were male, middle-aged government officials from Africa and Asia. I was impossibly and obviously young, female, and foreign. I wasn't even wearing a suit. Pulling back a chair next to Achala, I shrank myself down in the front row.

Pa Ousman leaned forward into his microphone and opened the meeting with a deep, "Colleagues, welcome to Durban!" Bubu took the seat beside him. "We have a full agenda to cover, but first allow me to give an overview of the negotiations that lie ahead."

The room quieted, and I listened to Pa Ousman with interest. As he spoke, people turned their country name plaques, which everyone kept referring to as flags, to stand vertically on end. When flags went up—the signal that a delegate wished to speak—Pa Ousman called on most people by name and answered their lengthy questions without referring to notes, unlike others who had small stacks compiled in front of them.

At certain questions, Pa Ousman and Bubu leaned away from their microphones to confer with each other in a practiced manner, tapping the other's arm when eye contact didn't draw the necessary attention. The physical opposite to Bubu's rotund figure, Pa Ousman sported rings that slid along slender fingers when he gestured, and his cheekbones stuck out prominently over a meticulously kept goatee. Perhaps as a result of looking over glasses, he appeared to draw himself inward, whereas Bubu's lips consistently parted in laughter, his hands rested on the belly he thumped at odd intervals for emphasis.

It was their pairing that caught my attention from the start. A good ten years between them, I guessed a familial relationship, such was the ease of their bond. When the chair called for a tea break, I decided to ask Bubu about it. "When did you and Pa Ousman meet?"

He tilted his head in thought. "We met at the department of water resources, a long time ago," Bubu answered, stirring sugar into his tea. "Pa Ousman still works there."

"Oh?" I knew there was more. Bubu's gray hair and open personality made me feel like I could ask him things. "So, you're not a government official, then? Have you retired?"

A laugh escaped him in a gust. He chuckled over it for a moment. "In The Gambia, retirement is a hole in the ground."

I waited, wondering if he would tell me more.

"I left the government to work as a consultant, but I will never stop working with Pa Ousman." He paused and held my gaze. "Good people are hard to find. You don't leave them once you do."

It sounded more like advice than an answer, so I tried to nod with the appropriate solemnity. But, really—*what?* My life was a tale of leaving to make a safer way for myself, Durban being yet another stop in a country where I knew absolutely no one, doing something I did not feel remotely qualified for. No risk, no reward. Right?

Over the next two days, Pa Ousman gave the floor to lead negotiators covering a plethora of issues: the latest Intergovernmental Panel on Climate Change outputs; taking forward the Bali Action Plan; and several other things that I would need to research in order to understand. The chair's technical knowledge came through in the questions he asked them, most of which were answered in the slow, moderated English I was beginning to associate with diplomacy. When delegates expressed differences of opinion, they accepted Pa Ousman's compromises with *hmms, tsks,* and nodding heads. Achala advised Pa Ousman by passing handwritten notes up to the top table. The chair often shared these with Bubu, and they would inform or remind the group in turn.

At one point, Pa Ousman tasked me with populating a list of coordinators. I didn't hold that post for long. The chair asked me to project a spreadsheet of vacant positions so that delegates could be volunteered in real time. Flags went up, and names from a variety of African and Asian countries sounded over the microphones.

"Mbaye from Senegal will lead the negotiation of market mechanisms."

"Fred Onduri of Uganda for technology transfer."

"For adaptation, Thinley from Bhutan and Hafij of Bangladesh."

Pleased with the volunteering, Pa Ousman signaled for me to start typing. After staring blankly at the keyboard and making wild guesses—*Mmbye from Senegal, Fred Undery from Uganda*—I was relieved of the task. Spelling has never been my strength.

I took copious notes, mostly to improve my minimal understanding of the UN climate negotiations. Otherwise, I edited documents, distributed publications, and generally tried to be helpful to my fellow Gambian delegates. I found that I didn't need much instruction to complete the assigned tasks and that people were friendly enough. But, in a room full of men, eyes were following me too closely to be noticing just my work and I wasn't getting a lot of *eye* contact.

Though this made me uncomfortable, it wasn't like gawking was a new sensation. I grew up sticking out in the crowd. As small-town America's infamous minority—the mixed child born to an African American father and Caucasian mother—I was well practiced at walking into a room and ignoring the heads that turned for the wrong reasons. When I was a child, it seemed like everyone knew me but few wanted to know my name. Even within my own family, race relations were far from seamless. My mom's parents had disowned her when she married my father. Though my dad's dad had approved of the wedding, he insisted, "Mixed kids have no place." Firstborn, my arrival changed things as both sets of grandparents reconciled themselves with their lineage. I came up in the space between.

I looked much like my father. When people saw my family together, they recognized my mother's resemblance in my sister and his in me. Like his, my lips are full enough that when I laugh, they press against my bottom teeth, breaking the circular O with two pink peaks. I have his toes and fingers, long and thin with dark lines of pigment running through the nail. I also have his well-defined calves that people associate with running and his comically small ears, the right of which is pointed at the top. My eyes are brown, except in the directness of the summer sun. The green hiding in their depths is the most recognizable nod to Mom, whose eyes are blue. I had always tended toward voluptuous. In school, I could never pass the fitness test, not because I couldn't run fast enough or do enough pull-ups but because I weighed too much.

As an adult, I sought out places where difference would not so define my experience. I won a travel scholarship at twenty-one and happily ran away. Free and alone, I traveled for a year. In Europe, I found race dynamics similar enough to know my place. In Brazil, I was delighted to blend in so seamlessly that the beautifully mixed thought me native. In China, people stared, though not in the sexually frustrated way they tended to in India. It felt more like unrestrainable curiosity, as if they were witnessing alien life and just could not help themselves.

Chinese citizens worked together to take my picture without my knowing. One minute I would be walking down the street or sitting on a park bench and suddenly someone making a peace sign would be right next to me smiling intently at a friend's camera. I desperately wished to see those pictures, betting my expression was priceless. The only people who appeared to share my experience were the North Carolinian students I ran into at the Beijing airport on the day I left. One was wearing a T-shirt whose characters read, WHAT THE FUCK ARE YOU STARING AT? I laughed aloud in the terminal as they passed.

Being different was certainly not a new experience. Having this difference be desirable sure was.

In the mingling that followed the close of the LDC Group's preparatory meeting on Wednesday, several negotiators extended drinks invitations. I politely declined the first and pulled my hand out of the crushing handshake of the second with less grace and more directness. By the time the third offered me livestock, my patience had run dry. "You're worth fifty-two goats," he said confidently, after obvious assessment.

"Excuse me?"

"Fifty-two goats. Your father will approve."

I laughed.

Hard.

It wasn't nice, but it couldn't be helped. I knew we came from different cultures; I just couldn't imagine a context wherein valuing someone in livestock was an acceptable greeting. Ick. He didn't look amused. Thankfully, Pa Ousman and Bubu were crossing the emptying room, so I moved to intercept them.

When I made it over, Pa Ousman asked, "Brianna, please can you send the presentations to the group? And the updated list of coordinators?"

A speedy task-completer, I made a good teammate in a quick-turnaround environment. "Already done," I said. "I've also sent you and Achala the action points I noted down over the past two days."

Pa Ousman raised his eyebrows in slight surprise. Bubu waved me through the door ahead of him. "The Gambia always has the best delegation," he joked to Pa Ousman. Approval noted, I smiled with relief at another day's work complete.

On Thursday, Achala introduced me to a colleague of hers who had arrived for the start of the negotiations. "This is Marika, my right-hand woman," she said as a slender, freckled personality with fading auburn hair approached our breakfast table. They hugged, and Marika shook my hand with a wide smile.

"Lovely to meet you," she said in a pronounced British accent.

Returning her smile was easy. I loved a British accent. I blamed Jane Austen and the plethora of British programming featured on PBS. Growing up an Anglophile was the obvious choice.

"I have a slightly odd favor to ask you," Marika said after Achala excused herself.

Marika's role involved distributing money to the LDC negotiators whose participation was supported by the International Institute for Environment and Development. But IIED's travel insurance policy didn't allow a single employee to carry thousands of dollars across borders. Nor was Marika especially thrilled about withdrawing that much cash alone. I was happy to go with her.

On the way, Marika offered plenty of conversation about the well-thumbed South African guidebook she carried in one of her three handbags. When we arrived at a central bank branch, I thought the teller's insistence that her stack of traveler's checks be numerically ordered a ploy for getting rid of us. Marika simply pulled up a chair opposite his station and started sorting. It took us several hours to complete the transaction, by which point the teller had grown so concerned that two young women would be walking the streets that loaded with cash that he drove us to the convention center himself. We had spent so much time together by then that the act felt perfectly natural.

We found Achala and a handful of LDC negotiators in the now set-up LDC office. The windowless room allotted to the group held a row of desktop computers on tables lining one wall and a round table in its central space; its adjoining conference room locked with a key. I cleared a place from cluttered papers at the round table and settled into a chair. Achala led Marika around the room introducing her to Pa Ousman, Bubu, and the other negotiators present.

"So, you're the one with the money," I overheard from across the table.

I looked up to see Marika shake hands with the delegate who had introduced himself to me as Dr. Y on the first day of the LDC Group's preparatory meeting. Even as a grad student, I met very few people who introduced themselves as "Doctor." It certainly made an impression.

"Please confirm Saturday's meeting," Pa Ousman interrupted my eavesdropping. I nodded, mourning the weekend. Apparently, Saturday and Sunday were working days here.

I spent all of Friday editing documents and sending out emails in the LDC office. When Marika leaned over to ask, "Want to eat out tonight?" her question broke the quiet drum of typing that had settled over the afternoon.

"Some of my IIED colleagues are going to the Chinese restaurant down the beach from our hotel if you're interested," she said, smiling. Marika was new to IIED, and her eagerness to meet the other members of staff was catching.

"Sounds good," I answered. It was a smart idea to get to know as many people as possible before the negotiations started on Monday.

Later that evening, Marika spotted the IIED team at a huge circular serving bar when we arrived at the restaurant. A chef stood chopping things in its center. Marika was in the middle of introducing me to people when a cell phone ring interrupted her. "Oh sorry, it's someone from the LDC Group."

We paused the introductions to let her take the call.

"It's Dr. Y," she mouthed after a second.

I gave her a sympathetic look and claimed a chair. From what I overheard of Marika's conversation, I gathered that Dr. Y was asking for my number. She gave it to him while looking at me commiseratingly, so I stepped outside to take the call we all knew was coming, wondering if I had forgotten to send him PowerPoint slides or add his email to the correct mailing list.

My phone vibrated. "Hello?"

"Hello, Brianna."

"Hi, Dr. Y. How can I help you?" I dug around in my bag searching for a pen and a piece of paper, hoping his request could be dealt with later. I didn't have my laptop on me.

"I'm nearly back to my hotel," he said.

"I'm sorry?"

"I have to take the bus to a shopping mall and then wait for a shuttle from my guesthouse to pick me up."

I paused, waiting. Transit routes were not where I thought this conversation was going. "OK?" I let the confusion sound in my voice.

"What are you doing now?"

"I'm out to dinner," I said, not understanding the small talk. "Was there something you needed?"

"Come to my hotel," he said.

"What?"

"I want you to come to my hotel."

The word *why* was forming on my lips when I started to guess at the meaning of his call and that actually asking why would elicit an answer I didn't want to hear. This wasn't about work at all.

Dr. Y was old enough to be my father. During my days in Durban, I had spoken maybe a handful of words to him. I knew nothing about him, nor he about me. The persistence of men in South Africa was proving slightly overwhelming. They crossed the hotel breakfast room to leave me business cards with cell numbers underlined for emphasis, hopped out of cars that hadn't come to a complete stop, walked up to me in restaurants.

A silence stretched the line, until the only thing I could think to say came out.

"No."

"What are you doing tomorrow?" he continued, unfazed.

This wasn't some guy on the street or from the packed meeting a few days ago, the vast majority of whom I hadn't seen since. Dr. Y and I were meant to work together. I tried to figure out how to end the conversation, while also reiterating what our relationship was supposed to be about.

"The LDC Group doesn't have any meetings tomorrow," I said. Then my thumb pressed down on the red button, ending the call.

Back in the restaurant, everyone was talking across the serving bar just as before. I reclaimed my seat and noticed that, in my absence, Marika had started piling food on my plate. The kindness of this struck me. I picked up a set of chopsticks and held them for a while, listening to the reunited colleagues continue their conversation with voices raised over the noise of the sautéing chef. I felt numb, like I couldn't feel my feet.

"What did Dr. Y want?" Marika asked from the next chair.

I raised my head to look at her, scrambling to think of something normal he could have said, to avoid embarrassment. But her eyes were soft, and surprisingly, the truth came out instead.

Marika's eyebrows crinkled together in stages of understanding: confusion, realization, indignation. "What? That was a booty call?!"

"Seems like it," I answered. "I can't think of any other reason why I would need to go to his hotel." The fact of this hit me.

"That's disgusting. I'm so sorry! I should never have given him your number. I swear I thought it was about work." Her words tumbled out in mortification.

"Of course you did. I did too."

I looked down at the noodles on my plate.

"I'm so sorry." She rubbed her hand across my shoulders. "What a creep."

I exhaled a laugh, feeling better. She was right. He was a creep. I didn't like that he had my number and I couldn't fathom the directness of a man who called unsolicited, but it was over. Marika's colleagues were laughing about something, and the food looked good. There was no reason to let him ruin my entire night.

Pa Ousman and Achala weren't in the LDC office when I walked in the next morning. I had stayed out with Marika and crew longer than anticipated, enjoying their laughing company and the lazy beachside stroll back to the hotel. The emails I was meant to send hadn't quite gone out yet, and I had missed both Achala and Marika in my rush to the conference center—choosing extra minutes in bed over breakfast. When time limited my options, sleep always trumped the morning meal.

Just inside the door, I stood adrift now, contemplating who at the central table might know where I could find the chair. I still wasn't entirely sure what I was meant to be doing, apart from delivering Pa Ousman's and Achala's ad hoc requests. I was lost without them.

"What are you doing tonight?"

I hadn't noticed Dr. Y when I came in. I certainly hadn't clocked that he was now standing too close to my left shoulder. In my knee-jerk side-eye, the certainty on his face turned my stomach. Gross. And I suddenly regretted skipping breakfast. Intent on ignoring him, I raised my chin and tried to step right around him. I didn't want to have this conversation. What I did in the evening was none of his business.

In response, he placed himself directly in front of me. Six feet tall and immovable, he stared into my face, blocking my path. If I wanted to get farther into the LDC office, I would have to go through him.

And now I wasn't just annoyed and uncomfortable.

Now I was scared.

In my experience, men who used their bodies rather than their words were to be feared. And that fear, of the violence and the pain I'd learned to expect, froze me to the spot. *Please, just go away.* I spiraled, locked inside my head.

"What's wrong with you?" he demanded in a voice too loud for the space. Everyone turned, the room quieting as they clued into the scene. All the people in the office were men. And now they were all staring at me.

Fuck.

I turned on my heels, ran out of that office. What was wrong with *me*? Before conscious thought kicked in, my feet carried me downstairs, around a corner, and into the nearest bathroom. The fluorescent lighting of the ladies' room bounced oddly off the tile floor and illuminated a woman washing her hands at the mirror. I felt safer already. Inside a stall, my forehead pressed against the closed door before I started to cry. An ocean and a hemisphere away from everyone and every safety net I knew, I was enveloped in the depth of my vulnerability. It felt like the world tipped and left me to slide untethered. The few strings of "known" that I was hanging on to were suddenly gone. I thought hanging up on him last night would be clear enough. My literal running away probably wouldn't faze him either.

When will he stop? What if he doesn't stop?

I panicked.

There would be no avoiding Dr. Y for the next several weeks. He would be in all the meetings and would apparently be spending time in the office. How the hell was I going to go back in there? I closed the toilet lid so I could sit with my head in my hands, crying until the toilet paper ran low.

When I could speak unwaveringly, I called Achala to ask her exact location rather than risk wandering around looking. I went straight there, intent on not missing more than I already had, and spent the day shadowing her and Pa Ousman, typing robotically for as long as was necessary. I flaked out of dinner, claiming I needed to finish some schoolwork, and headed for the hotel. Under a gathering storm, the Wi-Fi wasn't quite stretching to the nineteenth floor, so I huddled in the empty breakfast room to gobble up messages from friends and family. I scrolled through Facebook, tried to remind myself of my place in the world, tried to read, tried to think about anything else.

But I couldn't.

Instead I stared at angry clouds rolling in over the Indian Ocean, arms around my knees. The sun set, and the rains came in a torrential glut that kicked up waves. Gusts rattled the windows in their frames. And still I watched, trying not to think.

Around 9:00 PM, uniformed hotel employees started congregating across the room, arranging chairs into rows. I happily ignored them, and rather than move, I intently watched a cockroach crawl along the wall. Eventually, a staff meeting kicked off with congratulations for surpassing the hotel's monthly guest quota. Fully prepared to ignore the rest, I clicked the touch pad of my sleeping laptop and started an email.

A voice sounded with distinction, drawing my attention away. "As you know, the COP conference will be in Durban next week." My eyes refocused, taking in a manager.

"Now, what does COP stand for?" he asked, pausing for effect.

None of the staff seemed to know.

"It is not mean policeman," he joked, and the corner of my mouth twitched upward. Several staff members leaned forward attentively. "No. COP stands for the Conference of the People."

By now the manager had earned my complete attention. I tried not to revel too much in the fact that I knew better: COP stood for Conference of the Parties. I turned in my chair to face him.

"They are here in Durban to discuss climate change. And what means *climate change*?" he went on. I could tell this man liked holding a room's attention. He let the question hang unanswered before responding to it himself. "Climate change means that in fifty years it will be so hot our children will have to walk around in space suits."

My mouth popped open.

Several staff members nodded very seriously.

"The grass will die so we won't be able to graze our cattle and goats, and the ocean will reach reception," he said, building in emphasis. Then he continued quietly, "So, they are here to do a very good work for us, for our children, and the world. A very important work."

I sat in wonder looking at the gathered hotel staff, shocked into mindfulness about where, in fact, I was and the potential of what was about to happen.

People started filing out around 11:00 PM, once the meeting finished, making way for cleaners wielding industrial-sized vacuums. I gathered my things and headed upstairs to my room. Once showered and pajamaed, I shook my hair out into its protective sphere and crossed the room in darkness, slipping between the curtains and the floor-to-ceiling glass. The deluge of rain outside carried the scents of sea spray, sewage, and wet tarmac through the cracked window in gusts. The wind snapped the fabric behind me against its rail and raised goosebumps across my shoulders. Yet this seemed the safest place, cocooned here against the torrent.

I felt like a complete and total idiot. Not even a full week in South Africa, and all I wanted to do was leave. The negotiations hadn't even started yet!

I unwillingly watched the autoplay of the day's events run through my mind again, defeated. Of all my envisioned missteps, not sleeping with someone wasn't one of them. This wasn't what I signed up for, what the UN was supposed to be like. If I wanted sleepless nights, I could have stayed in my father's house. I had forgotten why I came to Durban in the first place, why exactly I thought it so important. Though earlier I had smiled at the hotel manager's small mistakes, his thesis was perfect.

My country talked about climate change like a minor risk it could afford to ignore. Americans granted endless airtime to debating the issue's legitimacy, discussing the economic cost of acting as though inaction was a viable option. Most of the world did not enjoy such luxury. Their newscasters didn't talk about climate change in economic figures, infrastructure faults, or the rising cost of flood insurance. They listed the names of the dead, put up images of the people who would no longer be able to feed themselves, children without homes to return to.

The governments I came here to work for understood this. One billion people lived in the Least Developed Countries. More than a third survived on less than two dollars a day. More than half did not have access to electricity. Together, they emitted less than 1 percent of the greenhouse gas emissions responsible for the climate crisis. The vast majority were subsistence farmers, fishers, and pastoralists. One bad season, one bad harvest, one natural disaster could end their entire families.

South African hotel workers knew that. As did I.

But knowing that and staying here meant different things.

Too long in staring, my vision blurred, catching my own reflection in the pane. My oval face was angular when captured amid strong emotion. It looked soft and vulnerable now. Though I traced my features— the shape of my brows, the set of my lips—I couldn't name the emotion underlit in orange streetlight. I could stay, fake an illness, wait for my fellow students to arrive before braving the conference center again.

Or I could go.

Run.

Cross the world to some other destination, far from anyone I had ever known. Start again. Go back to Brown, call this a learning experience. Try some other opportunity of the Ivy League. The eyes mirrored back to me were hollowed with indecision.

Lost in options.

Afraid of resolution.

Stumbling and unsure.

2

CLIMATE CHANGE AND ME

Seattle, Washington
March 2006

My AWARENESS OF CLIMATE CHANGE began with a feeling. I can think of no better way to describe it. It happened spring quarter, freshman year at the University of Washington.

I had moved to Seattle, the Emerald City, elated in the fall of 2005. It had garnered renewed fame from the ubiquitous popularity of the TV show *Grey's Anatomy*, which was airing its first enticing season. The largest city of the Pacific Northwest, Seattle's metropolitan area was surrounded by water and mountain ranges. They forged downtown in a narrow jumble of skyscrapers that ran right up to Puget Sound's waterfront, where people threw fish in Pike Place Market and boarded ferries to outlying islands or watched cargo ships come into dock.

The Seattle of my childhood was an old friend, full of new and exciting things, where, if we were good, Granddad would take us to the alien-like Space Needle. Or to see the urchins, starfish, and sea otters of the aquarium. Or to the Columbia Center tower's viewing platform where we could feel seventy-six floors sway beneath our feet. We biked the city's trails to reach Ivar's famous salmon chowder, played in count-less parks of coniferous trees—most always under the protection of a gray sky.

Home to a student body of forty thousand, the university occupied a peninsula north of the city center. UW's leafy campus hummed with collegiate gothic buildings. Brick and stone were a rarity in my experience of the Northwest. Walking through the quads and squares made me feel like a grand scholar released into a wide world of possibility, a feeling helped by the sweeping views of Mount Rainier, which, on clear days, stood towering and snowcapped in the distance.

I busied myself with discovering the joys of bubble tea and crashing frat parties with my dormmates and subsequent new best friends. We lived three to a room, crammed together in McCarty Hall, strangers who would share everything our first year away from home. I came on scholarship and drank in the freedom of living how I wished. The enjoyment of it caught me at odd moments—smiling into the mirror of the communal bathroom, sleeping more soundly than I ever had—so soundly that I grew an inch.

Six months in, I thought some buckling down was necessary. My freshman schedule was surprisingly empty compared to the long hours of high school. Counselors assured me that "prearchitecture" would keep me plenty busy in the years to come. One of my earliest memories was seeing my parents' model of our house made entirely of Popsicle sticks. It looked so fun, like a dollhouse. Students needed to apply for the architectural studies major at the end of their sophomore year with good grades in a full range of complementary electives.

"And I want to graduate with honors," I insisted.

My parents had made it abundantly clear that, after undergrad, I was on my own. Doing well in the classroom was infinitely more important than enjoying the university experience. I didn't really understand how, but adults were constantly assuring us that Latin labels on a diploma connected the bearer to a steady income.

"Well then, you'll need to attempt a science this year too," my counselor advised.

Taking the Introduction to Environmental Studies course felt like the easiest way to go. The first lecture had me sitting in one of the two hundred or so uncomfortable wooden chairs that filled an auditorium like so many others on campus. A professor took the podium and switched on the projector. The standard hush fell. The studious among us put pen to paper, the less so looked up from Facebook. I opened my notebook and wrote *Environmental Studies* across a page. I wondered which topic we would tackle first. Maybe oil spills or smog? We could dive into pesticide use?

"The majority of this class will cover climate change," the professor said while his PowerPoint presentation loaded.

I wrote the words *climate change* down with a curious expression. I didn't know what they meant. He continued as though everyone was familiar with the phrase, clicking over to a picture of the Earth from space.

"Carbon dioxide, methane, and other gases humans produce enhance the natural greenhouse effect of the Earth's atmosphere. The rising global average temperature this causes has varied effects on the Earth's systems and environments. The impacts of this climate change will be the focus of this course."

What?

The professor flipped over to a slide that looked like a respiratory graph. "Scientists began measuring the concentration of greenhouse gases in the atmosphere in the 1960s," he went on, not stopping.

Wait, what?!

I wanted to put my hand up and ask for a moment to process this information. People were fundamentally altering the Earth's climate, and this *wasn't* breaking news? All the other students appeared so terrifyingly unperturbed. The professor calmly explained how scientists came to identify the problem before I was born. How nations had negotiated entire international treaties and—in the case of the United States—abandoned them for well over a decade. This he said in dull tones like they were facts everyone knew, before advancing to another slide.

"The increase in average global temperature and the resulting change in climate will bring about ice melt in the poles," he said. *The slide behind him pictured Alaska, looking brown as dirt.*

"As well as varying weather patterns." *Slides of snowstorms in Georgia and 110-degree heat in Texas.*

"Ocean acidification." *A slide of bleached coral reefs and clams with deformed shells.*

"Prolonged periods of drought." *Slide of an empty California water reservoir and a raging wildfire.*

"Sea level rise and subsequent groundwater salinization." *Slide of a Florida suburb overrun with salt water.*

"Increased ocean temperatures that fuel more powerful hurricanes and tropical storms." *An aerial shot of Hurricane Katrina barreling toward New Orleans, six months ago.*

No! screamed through my head. I wanted to push the vortex off course, protect them. I didn't want to see the images that flooded my mind, now irrevocably tied to the word *Katrina*, the people drowning in their attics, water up to their necks. No way out. The cornrowed kids on rooftops. Grandmas in wheelchairs, waiting for help that wasn't coming. Thousands left to die. We fueled that monster? It couldn't be. We jacked that storm?

"Increased ocean temperatures will also affect rain and snowfall patterns." *Slide of barren Oregon ski slopes and a flooded midwestern town.*

"Which will exacerbate varying weather patterns and change the length and time of seasons." *Slide of a budding tree covered in frost.*

We were changing the length and timing of the seasons?

"Contribute to biodiversity loss and extinction." *Slides of dead salmon too warm to spawn and geese migrating to a barren plain.*

"Put additional stress on agriculture and global food security." *Shot of dead corn and the ribs of a famine-stricken child.*

What the actual fuck?

"And fuel resource-driven conflict." *An image of rifles in the desert.*

I caught myself gripping the foldout table of my auditorium chair, clammy palms to the surface of the wood, as I breathed in shallow pants. Not again. Out of *that* house, I thought I'd be safe. Out of *that* house, I thought I'd be free, the whole world ahead of me to build a future that was my own—stable, certain. I couldn't still be getting the shit kicked out of me. I couldn't still live in fear, be left for dead in a gathering storm, no other home to go to. Was nowhere safe?

Was nowhere fucking safe?!

And suddenly I wasn't in the lecture hall anymore. The wood was still slippery under my hands. The air still whistled in my ears. But it was another table. Another sentence, another day. The last time I gripped a tabletop, Dad was home that weekend. He left his job at the local hospital when I was in the fifth grade. Mom kept hers. He took another across the country and said we would join him in Virginia when the house sold.

Only it didn't.

The house my parents built didn't elicit broad appeal. It sat at the apex of their five-acre property, the highest point of the immediate landscape. From the outside, a glass wall dominated the facade. It was topped with a cacophony of slanting roofs and bordered by cedar decks, the largest of which was shaded by an immense, flowering wisteria. The building was large, even by American standards, and the doorbell's orange light glowed beneath the clouded sky.

The structure found its distinction in design. Old as time, passive solar buildings were all the rage when my parents took their plans to an architect in the early 1980s. Simply put, our house used the sun to its advantage. A wall of south-facing glass warmed the thermal mass of a two-storied brick hearth in the winter and—with the help of a thermal shade—cooled it in the summer. The rest of the house surrounded this central space in various rooms for my parents, my sister, and me. The only one not to benefit was our outdoor cat, Max, who preferred sleeping in the woodshed. Then again, there would be no need for such a space without the massive fireplace.

The community that lived along our gravel road was a close one. Outside city limits, the two dozen mailboxes of its property owners marked the road's turnoff from the paved street maintained by the county. Property sizes were measured in multiples of five acres. Our neighbors had horses and cows. I saw herds of elk and the occasional bobcat and black bear. I grew up knowing the comings and goings of each family, who got their lawnmower fixed by whom, who had seen the dentist's daughter the other day and where.

My parents were one of several families that built their houses in quick succession. They forged, as they called it, a gang in solidarity. My sister and I were among the group of school-aged kids who grew up in its ranks. There were barbecues in the summer, fireworks on the

Fourth of July, and the annual neighborhood Christmas party where I happily sang carols and gawked at the twenty-four-foot tree covered in lights. I dreaded being stuck with the "road repair kit" during the white elephant gift exchange. The kit was a bucket and a shovel. Should anyone be unlucky enough to pick it, they also earned a round of applause.

Filling potholes was the worst, especially in the snow, which, at one thousand feet up from the valley floor, we saw most winters—even if the snow line extended only to the end of the road. From the hill's west side, where the rock quarry cut a tremendous view through the woods of the valley below, the lights of my hometown were reflected in the curve of rivers. Kelso, Washington, population: twelve thousand.

Local news headlines at the time were dominated by the slow-moving destruction of a landslide one hill over from ours. The Pacific Northwest had seen years of above-average rainfall—a truly impressive amount, even for us. Combined with increased development, road excavation, and logging operations, the clay of the neighboring ridge had begun to move. Pictures of twisted roads and homes thrown off-kilter as they were swallowed by the earth covered front pages day after day. No matter how many surveys stated the geological credentials of our house, no one was buying.

A year passed without Dad, a year of waiting for the other shoe to drop, for the house to sell. For us to close the half-packed cardboard boxes and move east. A year of not wanting to leave my friends. Not wanting any of it.

A year we got to spend with just Mom, a native Ohioan, graduate of *The* Ohio State University who still made peanut butter–chocolate buckeye candies every Christmas. When she talked about her looks, she blamed her family of German immigrants for her broad hips and shoulders. She was happiest in her sewing room and in the garden, where she grew everything from flowers to pears. The branches of her favorite tree became so overloaded with apples every August that supports were needed to keep them from breaking. I ate her homemade blueberry jam with a spoon when she wasn't looking.

She wore trifocals, and her anger never lasted for long. Chanteal and I knew that if we played our cards right, we could usually get her decisions to swing our way. Our most ridiculous stunt involved serenading her in public places. We were taught the song "Wind Beneath My Wings" in grade school choir. It was the perfect thing to sing when she tried to refuse us.

"No Cap'n Crunch this week," she said, angling the shopping cart farther down the grocery aisle. In our house, sugary breakfast cereal was a privilege, not a right. It was earned through exceeding expectations—or begging.

I shuffled forward, clearing my throat.

"It must have been cold there in my shadow. . ." I sang quietly. Chanteal smiled. Then her face crinkled in a desperate struggle to remember the words. She would back me up once I hit the chorus. "You were content to let me shine." I paused to give my sister a meaningful look.

"That's your way," Chanteal beamed at my mother, remembering now.

Mom deliberately pushed the cart forward, trying to ignore us.

We weren't deterred.

"Did you ever know that you're my hero?!" her twelve- and ten-year-olds belted out. The family at the end of the aisle looked over.

"Stop it!" Mom laughed, unable to hold it together any longer. "You're going to make me pee my pants!"

"For you are the wind beneath my wings," we finished proudly.

Shaking with laughter, Mom pulled a handkerchief out of her pocket to dab the moisture in the corner of her eyes. Red in the face, she just nodded when we threw the cereal boxes into the cart.

I was in the sixth grade when Dad had to come back. His presence, once again, filled the entire house and determined what was said and done. I had changed school districts and gained an entirely new set of friends. Had my first boyfriend. In giggling sleepovers, I discussed every aspect of my budding teenage life: what was cool and what wasn't, what to wear and what you shouldn't be caught dead in. I rode bikes over plywood jumps with the neighbor kids, divvied up teams for kickball games. Dad went away a lot looking for work, and when he came back to where our house remained unsold, his frustration with the lives we led was clear.

That weekend, he assembled us at the kitchen table to discuss a fight Chanteal and I had while he was away. I said, she said. We yelled. I slammed a door. Mom told us not to do it again and I thought the matter was solved. Four days later, I no longer remembered what the argument had been about. But Dad was keen to discuss it in detail, intent that we hadn't been punished enough. From my designated chair to his right, I listened to him recount Mom's telling.

"I thought we had instilled in you the value of self-control," he said to me.

I exhaled, surreptitiously watching him rotate a Bic ballpoint pen in his right hand. This was going badly. Dad fell on the opposite end of the spectrum from Mom, and earning his approval meant preparing yourself for a long war of countless casualties. He spoke in the decisive sentences of someone who believed he was right and whose judgments were final. An athletic five foot four, he spent a lot of his time building things in the woodshop at the back of our garage. Most of our furniture was his design.

"You slammed doors in my house?"

He had been talking for ten minutes, each sentence rising in volume. Conversing with Dad was difficult at the best of times. It wasn't like talking to other members of my family or my friends—where you could engage at will. Dad liked to be in control. Especially when angry, he decided what would be said in advance. Variations from this unknown script were not tolerated.

"What do you have to say for yourself?"

I tried to make eye contact with anyone else, but Mom and Chanteal were staring intently at the floor.

"Speak."

The desperate hope that saying nothing would suffice died quietly in my mind. Fear stuck in the back of my throat. My breath caught, and my hands started to sweat.

"I'm sorry." I tried to sound repentant. If pushed, I would not be able to articulate what exactly I was sorry for. Both the phrase and the tone were a knee-jerk response. To roll over and beg forgiveness was the safest course to de-escalation, what he most often wanted to hear.

I couldn't think of anything else to say to soften his gaze.

The silence stretched on.

"There will be no third incident of you slamming doors," he told me.

I didn't remember the first incident. Apparently, this was strike two. I scrambled to review the last six months, struggling to find when this happened before and dreading other consequences.

"What if there is?" The words escaped me without conscious thought. In the heartbeat of shocked silence, all six eyes refocused on my face.

Shit.

This was gonna hurt.

He hit me. Palm to the back of the head. Hard enough that I lurched forward in my seat, face to the tabletop. As usual, the pain of being struck registered first. Then the sickening motion between where my head used to be and where it was. My eyes opened and struggled to refocus on the wood grain inches from my nose.

"What did you say?" Close to my left eardrum. Each word distinct and incredulous.

Level with my sight line, I saw his ballpoint pen lying on the surface of the table. I felt nothing but grateful that he'd put it down. He didn't always drop it: the pen, the fork, the keys. Violence could make a weapon of anything, lodge it under your skin. The part of me that wasn't petrified of Bic-in-brain wanted to ask, *Will you hit me again if I repeat the question?* Most of me knew better than to say anything else. If I wanted the pain to stop, I would concede. I swallowed and kept motionless, prayed it was over.

The silence stretched on.

"That's what I thought." Dominance and the end of the matter. "Let's go."

I heard chairs squeak back from the table my clammy hands still clung to. It was a sunny Saturday afternoon in late summer. Of all places, we were supposed to be going to the fair. Slower than the rest, I rose and made my way toward the door to the garage. Mom stood just inside it.

She was crying.

"We aren't going," she said.

"What?" My father's voice registered more faintly. He must have already been in the car.

"Why is it that every time we sit down to discuss a problem, you have to haul off and hit somebody?" Mom yelled.

I stood behind her, shocked. I had never seen my mother openly defy my father. Ever. In that house, I was the only one stupid enough to mouth off, to say anything. And I, certainly, was the only one who got hit over it.

"Chanteal, get out of the car," Mom went on. "We're not going. How can you expect them to promise that they'll never argue again?"

And I lost it. Bawled myself blind. Mom had yelled at Dad. Mom had stood up for me. By the time I stopped, my father and the car were gone, and Mom was bending down to hug me. "I love you," she murmured into my shoulder. It registered differently than the other times she said it, which were usually in parting. Or on special occasions. This time, I understood it. Love as a verb, not a benediction. An act of protection.

She didn't say anything more about it. We never talked about conflict. The expectation was to say nothing and move on. I followed my sister to the stereo room, where she turned on an Indiana Jones movie. While she watched it, I stared out the window, anxiously wondering if Dad would come back. If he didn't, it would be my fault. My parents never fought—not like that. Theirs was a silent war. Dad said and did things Mom didn't like, but her response was to stay quiet, at least around us. To hide. I was surprised that Dad was home by dinnertime. I was not surprised that when we sat around the same table to eat, we pretended like nothing had happened.

Despite appearances, our house was silent a lot.

My middle school teachers praised Dad following their parent-teacher conferences. Most of my friends' parents too when they first met. His wit and athleticism inspired admiration. They saw his intelligence, and his features, in me. And I wouldn't correct them. The silence in our home lingered in me as I followed whosever parent had come to collect me out to their car. Since our house wasn't on the way to anything—all destinations were a minimum drive of twenty minutes—going to town before the age of sixteen meant getting a ride. Though the parents were perplexed as to why I needed a lift when an adult opened the door, Dad's explanations were eloquent and they usually came away with a good impression.

I tried to imagine what they saw. I supposed it looked fine: middle-class family, big house, two-car garage, kids with braces and nice clothes. A father who worked as an ultrasound consultant for what he jokingly called "the pretty boy clinic"—a private establishment in Portland, Oregon, where he spent days scanning varicose veins for people rich enough to shrink them—or weeks away working the Seattle hospital circuit. As time passed, though, my friends' parents would eventually start asking questions:

"Why is your house always dark when I come by these days?"

"Does your dad ever come to school events?"

"Is everything OK, dear? This is the second time I've dropped you off around dinner and no one's been home."

At fourteen, I shrugged and searched for my keys. "My parents work a lot," I said, not knowing where they were but positive that they weren't together.

"Do you need a ride to school tomorrow?" was a common follow-up question.

"No, that's OK. I can get a ride with the neighbors or get the bus from the mailboxes at the start of the road."

"If you're sure."

"Yup, I do it all the time. Getting to school is not a problem."

Alone in the driveway, I would wave them off—holding a smile until their taillights disappeared.

In truth, I loved the refuge of school. My middle school was directly across the parking lot from Kelso's only public high school, which turned out a graduating class of about four hundred. I was one of the handful of Black students that made up a 2 percent minority. Grassy athletic fields, a football stadium, and baseball diamonds surrounded both schools, and they all tucked neatly into the curve of the Coweeman River. The high school even had a full-length swimming pool and a separate diving pool. I was an athletic nerd who liked people and was avoiding home. If the administrators had allowed it, I would have happily moved in.

In middle school I met the girl who would become my best friend, seated across from me in science class. During that first quarter together, Noelle and I only managed to impress on each other our growing unhappiness. When we met again as high school sophomores, one easy, entirely forgettable conversation kicked off decades of friendship.

Fashion standards embodied, Noelle was kind, funny, and effortlessly cool. She was a good listener, able to draw out conversation between two people without obvious similarities. Like the fact that she was tall, slender, blond-haired and blue-eyed. These were antonyms of my attributes. As my mother liked to say, "You two couldn't look more different." Mom was not the best for compliments on physical appearance. This was a trait inherited from her mother, who I once heard remark that Mom, trying on a sweater, looked like a tea cozy. "That does nothing for you" ended any idea of buying the button-down gray wool. They meant well, but Grandma and Mom did not sugarcoat. And it was true that Noelle's and my physical similarities were limited.

Noelle and I confided in each other as middle school crushes shifted into something more. We worked them out in giggles and notes, developed our own understanding of what to expect from the stories we heard. I had my first real kiss at fifteen. Noelle was the first to hear about it, waiting in Kelso High School's empty parking lot until I finished practice.

"So, how was hanging out with Daniel?" Noelle asked, stretching the "so" out into a question all its own.

I laughed, embarrassed as I told the story, and when I finally got to, "And then he kissed me," it was silent for about half a second before we both squealed in unison.

My days at Kelso High School started with a 6:45 AM elective zero period, prehomeroom, and lasted until the final whistles of sports practice around 5:00 PM. I tried everything: jazz band and weightlifting, soccer and student government, and track and the Knowledge Bowl quiz team. Even water polo for that one exhausting, early-morning season. But I preferred the classroom to the field. I liked to learn and had a knack for remembering details that made memorizing facts and figures relatively easy.

Even so, some subjects were more enjoyable than others. The fundamentals of science never clicked. Understanding how elements interacted was something I could memorize rather than something I knew. I wasn't good with languages either. The perfectionist in me was both too proud and too anal to enjoy sounding like a child over and over again.

I loved the release of history. My life was so marked by difference that I greeted learning about societies in times removed as a welcome escape. Modern events were harder to deal with. Perhaps I empathized with the people I read about in America's recent past differently than my classmates did. I certainly reacted differently to learning about them.

I read Upton Sinclair's *The Jungle* in a history class: 1905, in Chicago's meatpacking district. Immigrants forced into backbreaking labor in unheated slaughterhouses where it was difficult to see—conditions that killed off the old and maimed the young. Factory girls forced into prostitution. Deaths by food poisoning. A perfect storm of dangerous hazards and relentless cycles of poverty. Efforts to organize stymied by corporate greed and political corruption. Violence inherent to the system. Cruelty from which there was no escape—where everyone lived one false step away from death and disaster, and regardless of how hard they worked, they ended up worse off than they started.

The things Sinclair described had me filling my journals with questions, looking at the meals served to me with consternation. Where did it come from? Did this pork chop cost someone their arm? Their life? Things can't still be like this a hundred years later, right?

I found myself milling around libraries, combing through books in search of answers, not to a teacher's questions but to my own. I spent a summer reading books like *Fast Food Nation* and trying not to vomit, as I discovered why meat processing remained America's most dangerous job. The list of accident report titles was enough to send me reeling: Burned lungs from inhaling chlorine. Burned skin after fuel from a saw ignited. Eye injured when struck by a hanging hook. Hospitalized for arm amputation. Foggy work conditions that made it difficult to see. *How was that* still *happening? The workers can't* see?! Killed by an ammonia spill. Killed when an arm caught in a meat grinder. Decapitated by the chain of a hide puller machine. Killed by a stun gun. Killed when the head was crushed by a conveyor.

I took breaks when I started to feel sick.

That summer happened to coincide with the first time I saw my boyfriend's dad skin a deer. We were in the workspace of their detached garage, listening to rain hit the metal roof under a dark sky. A lamp shone on the layers of fur and fat he scraped back from a shockingly red lump of muscle.

"Maybe you shouldn't watch this," my boyfriend said from across the corpse.

I could not stop looking. I also couldn't keep my lips from pulling away from my teeth in disgust, a grimace untamed. I could gut a fish without thinking, but I wasn't prepared to peel skin from flesh, and now I knew that the people who were doing it held America's lowest-paying industrial jobs and would likely be fired before they passed the six-month mark, which mandated paid holidays and health insurance. I also knew that senators married board members of packing companies, corruption allowing employers to bypass workers' rights protections and enabling fatal greed and cruelty from which there was no escape.

And for what? Hamburgers?! Flipped by other teenagers in restaurant franchises where sexual harassment ran wild and most affected girls who looked just like me? It was horrible. No meal was worth that cost.

But what could I do? I toyed with eating less. I was always a "go big or go home" kind of person, though, and things I'd once found appetizing suddenly weren't now that I knew where they came from. For New Year's, I resolved to give up all meat but seafood in a move that left my grandma wondering how I would survive.

My mom's parents lived in an old farmhouse on the Ohio flatness inland of Lake Erie. The house, or the kitchen table, rather, where Grandma was born was less than a mile from every house my mom's childhood resided in. Their village was surrounded by dairy farm acreage and cornfields shadowed by water towers. Plastic American flags fluttered on tidy squares of green lawn outside white houses. Little girls stood watching the street from behind porch screen doors, sundresses around them. Everyone knew everyone, and all dared to live beneath a sky that shone blue above the graying cumulous of tornado country.

Mom moved away to escape the certainty of it all. The closeness—even being Kelso-born—was beyond anything I knew. Grandma and Grandpa's headstones already marked the local cemetery. All that remained unfinished were departure dates.

During our visits when I was in high school, Grandma baked apple pies in anticipation of our arrival, laughed with us over endless hands of cards, and added cream of chicken soup to dishes she told me were vegetarian. "It's just for flavor," she responded when I asked why the cheesy potatoes tasted of poultry. "You can have flavor, can't you?"

We locked eyes across the table, each exhausted with the other.

"I still don't understand why you're doing this," she said. We were in the kitchen after the meal, passing dishes through soapy and then clean water together before putting them on a rack to dry. "It's not like you need to slim down for a wedding, dear," she stated after finding my reasoning unconvincing. Grandma had the incredible ability to work marriage into every conversation we had had since I completed puberty. Where I saw fun in high school relationships, she saw great-grandchildren, a point she made abundantly clear. I found the singularity of her vision for my future exasperating beyond words.

Yet, talking about boys and boyfriends had become an increasingly important component of Noelle's and my conversations. Noelle's mom worked as a hairdresser, which made her house the perfect place to get ready for dances. The night of our first ball, we laughed through hours of hair and makeup before stepping into floor-length gowns. Our moms snapped pictures as we waited for the boys to arrive. In them, we're gripping hands so tightly that both sets look white. Noelle's date had managed to get his driver's license two days beforehand—an effort she had insisted on. We thought it the coolest thing that the four of us would ride together, no adults required.

I counted down the days until my sixteenth birthday, meticulously planning how closely I could align the date with a driver's test. I longed for the freedom of the car. My family seemed happy that I was finally old enough. Dad bought himself a new-to-him car, a Saab he drove like a maniac. Mom took me for my driver's test, shelled out gas money, and told me, "As long as Chanteal gets where she needs to be, you can go wherever you want."

Once I had my license, my days stretched into night games and curfew-testing dates as I mobilized my life out of Dad's old Toyota

4Runner. I spent most of my free time between ages sixteen and eighteen in either Noelle's car or mine laughing, crying, or gossiping in a
darkened parking lot. There was nothing we couldn't talk about.

Noelle was the first person I told about Dad, about my real relationship with him. Past his ability to put on a suit and go to work in
a hospital. Past the smile that charmed my teachers in parent-teacher
conferences. Past the pretending and the well-practiced silence. Past
the denial. And the lies. I told her about the man who hit me. Left me
crying. About the one who gave me nightmares.

She put her arm around my shoulders when the tears fell, and we sat
together in the silence that followed—two teenagers who knew nothing
more of the world than life in Kelso.

She listened to me.

She prayed for me.

And from then on, I knew there was nothing I couldn't tell Noelle.

───────────

I graduated high school valedictorian. When I left Kelso for the University
of Washington, I had no intention of coming back. Though I was intent
to study the efficiency of my parents' house and design other places that
used the sun to their advantage, I would never live in it again. Besides,
the metropolis I remembered from childhood was wonderful.

A hundred and twenty-five miles north from where Kelso's giant
yellow *K* pinned the interstate—a distance Dad could cover in less than
two hours—Seattle was where I always hoped we were going when my
parents told us to pack. It offered an intoxicatingly different rhythm of
life. We stayed at Granddad's house just north of Seward Park, walking
distance to the waters of Lake Washington. A retired paratrooper, Granddad lived alone in an undecorated, two-story rectangle of blue shingles.

I loved going there.

I loved the roses he grew out front by the wrought-iron bench. I
loved the "princess room" Chanteal and I shared—first door at the top
of the unfinished staircase—named for our insistence on stapling lace
trim to the empty doorframe rather than in reference to the decor. A
train set ran over sheets of plywood in the basement, and a wonky

piano filled the front room, whose walls were covered top to bottom in family photos.

Most of the pictures featured people I knew. After my grandparents divorced, Dad, a high school senior, followed Granddad out from North Carolina. They settled in the Emerald City, eventually attracting my dad's closest brother out west as well. My Uncle Charles and Aunt Vickie smiled together with the cousins who now lived in a neighboring suburb. Many, though, were the grinning portraits of the southern Crafts I rarely got to see, row after row of smiling faces I was related to but didn't quite know.

Well, most of the people in the photos were smiling. The exception was a candid shot of Dad. He must have been a teenager at the time; it was taken in Fayetteville. Dad stood on a basketball court wearing blue jeans and sporting a 'fro. It was sunny, and he held the ball between his left hand and his hip. He looked angry as shit. Mom laughed whenever she looked at that picture. I never understood why.

We went to Seattle in summer for the Seafair Weekend Festival, when boats raced across the lake. Granddad hosted a viewing party. Friends and neighbors came by before the Blue Angels fighter jets screamed in tight formation overhead. We lined the porch and the small balcony, squinting toward the sky until the thunder of their passing turned all heads in unison. People snapped photos of the blue and yellow planes twisting through the air in spirals. Granddad put hot links on the barbecue in the summer heat. My cousins ate the last of the peach cobbler, and Uncle Charles invariably tried to sell me on vanilla wafer banana pudding one last time.

"Taste it. You'll like it!" He moved closer with a spoonful of the stuff. Think banana-flavored custard topped with chalky cookies.

"Yuck." I laughed. "How can you eat that?" I stuck out my tongue and squirmed out of the kitchen, rounding the counter with such force that the pound cake wobbled under its glass lid.

"You're missing so much," he sighed, swallowing the spoonful himself.

"The ice cream we made is better." Then I noticed the cup in his hand. "Wait, is that my lemonade?" I shouted, laughing. "Give it back!" We reversed roles, chased becoming chaser. Everyone knew that if Uncle

Charles took a "sip" of your drink, that lasted until you forcibly removed it. But he was quick like Dad, and I couldn't catch him.

We ran out onto the front lawn.

"Look!" I cried, pointing in mock desperation. "Chris isn't eating his greens." Uncle Charles distracted, I won back the cup and earned a glare from my younger cousin, who had been successfully hiding the collards underneath his finished ribs all afternoon. I loved the loudness of the Crafts, where my assertiveness blended seamlessly into the boisterous dynamic of the extended family. When the festival finished, I watched with longing from the back seat while Dad loaded the car, gutted that he wouldn't let me accept Uncle Charles's invitation to stay longer. As a child, I never wanted to leave Seattle. As an undergrad, I didn't have to.

So here I was, gripping the foldout table of my auditorium chair, eyes stuck on calamities presented on the glowing screen in Introduction to Environmental Studies, doing all I could to impress the selection committee and accidentally stumbling into the epoch of my still-unformed existence.

It was all just so painfully simple. The greenhouse effect was hardly novel. If you hiked the concentration of greenhouse gases, its effects intensified. The American lifestyle produced carbon dioxide and other greenhouse gases—a tremendous amount of them. When I sat in lecture in 2006, both our annual and cumulative output was greater than any other nation on earth. So said the professor as he continued flipping through slide after slide of his unending PowerPoint presentation.

"Varying temperature patterns also directly impact human health," he said.

Oh God.

"Warm weather adds to thermal stress." *A picture of contractors bent double on radiating rooftops and farmers with heatstroke.*

"Increases the prevalence of allergens." *Slides of weeds in bloom and black mold.*

"Contributors to asthma." *Picture of Brown and Black children coughing in a smog-ridden cityscape.*

How was this actually and already happening?

"And the likelihood of cardiorespiratory disorders." *A graph tracking respiratory illnesses and heart disease against the duration of wildfires.* My eyebrows shot up.

"Warmer temperatures also fuel the spread of vector-borne diseases." *Slide of a mosquito carrying the words* WEST NILE VIRUS *and a tick beside the words* LYME DISEASE.

"The contamination of drinking water." *Aerial shot of algae blooming across the Great Lakes.*

"And the growth of viruses, parasites, and bacteria." *Kids swimming next to the words* SALMONELLA *and E.* COLI.

I couldn't concentrate anymore. I was too busy gasping. Blankly, I stared at graphic after graphic of climbing emissions scenarios, wondering when it would all end, lost in my airless panic.

And that was my first feeling of climate change.

After I managed to walk out of the environmental studies lecture, I found an empty bench on the quad. I sat there for a long time. Nothing in me doubted the reality of what I had been told. It was presented to me as fact, a phenomenon supported by over thirty years of scientific findings attributed to American and international researchers alike. Even with my high-school-level knowledge of science, the concepts were logical. What baffled me was the newness of it all. Why was I never told that the gases we polluted so frivolously were radically changing our atmosphere? How was a university lecture at age nineteen the first time I learned that people were causing the climate to change? And why weren't we doing anything about it?

I felt like I had been cheated out of understanding a critical piece of how the world worked. Drivers of the greenhouse effect should have been taught right along with the movement of tectonic plates, jet streams, and the other Earth systems we deemed so fundamental to a basic science education.

I loved my world, the mountains, and the sky, the twinkling, expanding city, and the lushness of the trees beyond. I had known nothing else, no other way of life. I thought about all the upheavals a climate crisis meant. I thought about how addressing the problem railed against the powers that be: the price of gasoline that so affected my day-to-day life, the means by which we generated electricity and fueled the American economy, the sprawling lifestyle I never thought to question except from an aesthetic lens. Suddenly the energy, the consumption, the expanse were more than a design question. Changing these systems, these vested interests, seemed so difficult, so unlikely—even given their lethal consequences.

Then there were the questions. I had so many questions. Of anger and fear. Loss and betrayal. Despair and retribution. Love and protection. And they all paled in comparison to the most important one, the one whose answers would change everything.

What should I do?

3

A CONFERENCE OF PARTIES

Durban, South Africa
December 2011

I SPENT SUNDAY COLLECTING MYSELF after that first bombarding week in Durban. The first step was to tip out my suitcase and rummage for the tennis shoes packed more in hope than expectation. I didn't consider myself a runner. Sprints were fun in high school—the adrenaline of rushing to cross the finish. I found distance running on most occasions little more than a long trial of convincing myself to keep moving. Now, it felt like the perfect thing to do.

That evening found me on the beachfront pedestrian boulevard striding through the fear, anger, and vulnerability. What set me sprinting was imagining what I would do when I saw Dr. Y again. *When*, because if I stayed, it was coming. Only leaving the country would prevent it, and that kind of running would also mean giving up the opportunity of a lifetime, the chance to understand how doing something about the climate crisis could become an actual career—one that shaped international decisions. That was the career I wanted but wouldn't have if I ran away before the negotiations even started.

After I had worn myself out enough to really think, I knew which choice I would most regret. Climate change was the single greatest threat faced by humanity. And those in the Least Developed Countries suffered first and worst in our changing world, though the problem was not of their making and would invariably impact us all—both those living and those to come. I wanted to stay and begin my own very important work.

When I told Marika about my run-in with Dr. Y in the LDC office, she was all sympathy and still mortified about accidentally facilitating his attempted booty call. Though it took me several stuttering breaths to get the story out, I felt relieved talking about it. My telling turned into an hour-long discussion of horror stories, which helped contextualize my first brush with workplace harassment. It was a big deal. But I wasn't the only one, and mercifully, mine was contact free. Spent and still a bit shaken, by Monday I was as ready as I could be.

In another set of my dwindling supply of UN-appropriate clothes, I went down to breakfast only to spot Achala dressed to the nines. She wore a loose-fitting cream-colored silk blouse tucked into a high-waisted wool skirt that was belted by a designer strip of leather. Heels again too. I had on the maid of honor dress that Noelle's mom had sewed for me for her wedding: deep blue with a jersey ruff across the shoulders, knee length, and beautiful, the touch of home I needed to come downstairs at all. And flats, dull black ballet flats.

Though I didn't want to draw attention to the contrast, I couldn't help telling her, "I love your outfit," in greeting.

"Thank you." Achala smiled. "Marika is sitting over by the windows."

After breakfast, I wondered what lay in store for the official start of the UN climate negotiations while a shuttle carried us away from the sand and surf bordering the hotel. A "Wow" from Marika brought me back to the present several minutes later. "They've really increased security since we passed this way on Saturday," she said, staring intently out the window.

We were driving down a nearly empty multilane road. A week in South Africa was enough to find this unusual. First I noticed the concrete roadblocks, perimeters, and chain-link fencing along the sidewalk, then the parked police cars and policemen sectioning off the street at its intersections. Security outside the convention center resembled trying to board an airplane. We had our bags run through X-ray machines, placed watches and belts in gray tubs, and passed through metal detectors. Guards patted people down.

The atmosphere felt much more serious than last week. I slipped my badge, the one with my picture and the word PARTY on it, over my head and a woman with a card reader scanned its barcode. "Welcome," she said after checking that the image on her computer screen was indeed me. I stepped outside, marveling at how much things had changed and squinting into the sun to locate Marika and Achala. The convention center's exterior no longer resembled the empty alleys of last week. It had transformed into something like a trade fair—only the vendors were stalls of environmental organizations pitching their research work.

"This is the zone for observers," Achala said after we regrouped. Marika and I followed when she started walking through the crowd. Nongovernmental organizations' representatives eagerly tried to pass us their publications or a branded USB stick. I got distracted taking in the names. All these people worked on climate change. The buzz of tents and booths in the observer zone stretched up to the open convention center doors.

Achala's voice drifted back to us over the din. "We need to find the area for Parties." Achala continued purposefully forward only to have a security guard appear, hand raised, outside the main doors.

He told her to wait, waving us toward yet another badge reader.

When PARTY flashed on its screen, we were allowed in. What a trip. I had never before worked in a place that involved so much security. When we were finally inside, the contrast with last week was palpable here too. The halls bustled. I turned around trying to get my bearings among the chatter. I recognized the corridors and meeting rooms, but I couldn't quite locate where in the convention center we were. Achala's cell phone rang again, and she answered in another language.

"The LDC office is upstairs and to the left, right?" Marika asked.

"I think so?" I answered. "It's hard to know because we came in a different way today." I would feel bad pointing her in the wrong direction.

"I'll find it. I have some expenses to sort out," Marika said. "See you both later."

Still on the phone, Achala craned her head and stepped closer to one of the TV monitors she had called a "live screen" when she pointed it out last week. I now saw why. The screen featured a scrolling schedule for the day, which also reminded me of an airport. Each line had a start time, the name of the meeting, and the room in which to find it. "Where's the opening?" she asked me.

We both watched until it came up.

"Plenary?" I read, unsure of the word.

"Good, let's go."

While walking, Achala occasionally stopped to chat with people, sometimes hanging up with whomever she was speaking to on the phone, sometimes not. Too busy taking in the variety of people also finding their way, I followed Achala through a line of double doors. I wasn't prepared. I had never been to a plenary session before.

Do you ever tear up during the Olympics? I did. Without fail, I needed a box of tissues to get through the opening ceremony's parade of nations. The beauty of the world united by a common purpose invariably overwhelmed me.

Walking into the plenary inspired the same emotion.

They were all there: every nation marked with a clear, white flag. The United Nations, assembled to stop the climate crisis. This was where governments agreed on international solutions to the global problem. I couldn't believe that I was here too.

Rows and rows of tables and chairs faced a stage that held a long table, all in a hall six times the size of the room the LDC Group had met in last week. It took us several minutes to locate The Gambia's designated four chairs toward the front of the alphabetically arranged room. Achala greeted Pa Ousman and Bubu as we approached, and we took our seats in the empty chairs behind them. I continued gazing around the room, ecstatic. I was about to witness the start of the seventeenth Conference of the Parties to the United Nations Framework Convention on Climate Change! With dilated eyes, I watched representatives from every country fill the hall, this realization flooring me.

With the globe assembled, a commanding woman with short hair in a multicolored dress ascended the stage. "She's the minister of foreign affairs, yes?" I heard Achala whisper to Pa Ousman. "The COP president?"

Minister Maite Nkoana-Mashabane was followed by a large procession of South Africans surrounding the man I recognized as their president, Jacob Zuma. As was custom, the host nation presided over the climate negotiations. The previous year, Mexico had overseen them. Now, South Africa would steer the talks. Success here could bring the prestige of naming an international treaty. Failure could mean diplomatic embarrassment for the foreseeable future. Such was the responsibility of hosting.

After an acknowledgment to the other heads of state present, Minister Nkoana-Mashabane shook hands with President Zuma and took to the podium. "It is an honor to preside over COP 17 on behalf of my people," her voice sounded over the microphone. "Durban is a decisive moment for the future of the multilateral rule-based regime which has evolved over many years. . . . We must work together to save tomorrow, today!"

I clapped with the crowd, noticing that delegates along the aisles were preparing to speak. Enraptured, I took note of the speeches. Their enthusiasm was everything I wanted to hear, from the people empowered to actually enact the changes that they promised. Australia, the European Union, Switzerland, they were all talking about further strengthening the international effort. Pa Ousman pushed the button at the base of the microphone in front of him to speak on behalf of the LDCs. He reminded the room that more than three decades had passed since the UN first identified the need for global mobilization to address climate change. "Recent findings are among the most convincing, highlighting not only the threats ahead of humanity—for which the most vulnerable countries are to bear the greatest burdens—but also the feasibility of innovative options as well as the opportunities to quickly shift into development pathways that limit the level of risk associated with a changing climate."

I didn't hear this kind of thing so plainly spoken in America. Here, it headlined the negotiations, and a world of governments stood ready to do something about it.

———————

When the speeches finished, I wandered toward the LDC office with trepidation. I knew confronting Dr. Y again was bound to happen. I just didn't want it to happen today, didn't want anything to tarnish the wondrous appeal of being in the moment. The office door squeaked on opening. Discovering that Marika was the only occupant filled me with relief. I exhaled, collecting myself enough to cross the room and sit next to her along the row of desktop computers.

"Is anyone going to the reception?" Marika asked the office after Pa Ousman and Achala joined. My earlier readthrough of the daily program informed me that, as was tradition, the COP presidency was throwing delegates a welcome party that night. I hadn't realized it was well past 5:00 PM. No one else seemed interested, though I was curious to see it.

Outside the convention center, walking down another road cleared of traffic by security perimeters, Marika shrugged at this. "They're old hats at these events. I'll bet Pa Ousman has lost count of how many COPs he's been to."

"Well, I'm glad we can go to our first COP reception together, dear." I beamed at her with overblown sincerity. I thought she would just laugh. Instead she wound her arm through mine, making googly eyes right back at me, and I giggled at the joke.

It was easy to be around Marika. I liked meeting anyone who could fill silence with laughter in under a week.

It was not going to be easy to get into this reception.

A long line of COP attendees snaked through the square bordering Durban City Hall, a grandiose building complete with colonial facade and dome. We located the end and settled in to wait. I took off my cardigan and stretched my arms out in the muted sunlight of the early evening. To pass the time, Marika and I exchanged questions about our lives across the pond. We shared cultural references and soon established enough mutual understanding to tease each other about the ridiculous aspects of our respective homes.

"And the queen, I assume you're on a first-name basis?" I asked. "How are the royals these days?"

"Marvelous," she said drawing out the -are-. "Thank you. Have you left your horse and gun back in the States?"

Again, I couldn't help but laugh.

The streetlamps came on, illuminating the line of people still only slowly moving inside. The square had plaques in tribute to famous South Africans laid into its tiles, like the stars along the Hollywood Walk of Fame. I recognized only a few. I smiled at Miriam Makeba's, who I knew as Mama Africa from a record my father used to play. Dad was a sucker for good music. He bought an electric baby grand for a house where no one played, just so we could properly tune the saxophones, clarinets, and trombones we did.

My stomach growled and I prayed we hadn't missed the food. Once Marika and I finally crossed the threshold, we made a beeline to the buffet. This we followed by grabbing drinks. I had just snagged us a place at a standing table when the stage lights started to flash. A speaker introduced one South African dance troupe after another. They performed to live music in a spectacular welcome to the motherland, and I craned my neck to catch the fusion of African high kicks and jumps with the detailed hand and footwork of Indian dance. I loved a dance; it told you so much about a place. The show was worth waiting for. The presidency had gone all out.

———————

Over the next few days, my mission became understanding what I was meant to be doing now that the negotiations had begun. Too early on Tuesday morning, I met with Pa Ousman in the LDC office to go over his meeting requests.

"Yes to the European Union at 2:00 PM. Yes to the South African presidency at 7:00 PM. Tell Bubu he'll need to chair the evening coordination meeting."

Armed with the login details Achala provided, I handled the corresponding logistical emails in the chair's account, updating his schedule as we went through the events of the day.

"No to the Indigenous Peoples' Forum. Ask if they can do tomorrow morning instead. . ." he went on between sips of tea.

I yawned and clicked the keys of my laptop.

The sheer number of commitments Pa Ousman agreed to amazed me. All told, he could easily fill a twelve-hour day with meetings back-to-back. Pa Ousman appeared to run on little sleep and the sack of bitter nuts he carried around with him. I marveled at how anyone could have something intelligent to say to diplomats, NGOs, and reporters from dawn to dusk.

The chair's workload was the reason I was here. The Gambia's twenty-one-person delegation included Achala, Marika, and me. The American delegation held ninety-five. Even though the world's poorest countries were the most affected by climate change, that didn't mean they could afford to send fleets to the UN to negotiate a solution. To bolster negotiating power, the outnumbered had to rely on bloc strength. Apart from having a hundred-man delegation, there was no other way one country could track everything that was happening.

I followed the gang to the LDC Group's coordination meetings, which gathered twice daily at 1:00 PM and 7:00 PM. Pa Ousman chaired as group members updated each other on the progress of the issues they followed. I compiled the agenda and dealt with communicating with any guests, the non-LDC delegates invited to attend. Sometimes scientists gave presentations on the latest climate data. Sometimes the COP president visited to brief the group on progress and ask advice on unsticking tricky issues.

After the meetings, I added email addresses to the various lists the group maintained so that delegates following the same topics could keep in touch. I also sent out the presentations guests made and fired off notes to those who couldn't make the meeting, including Dr. Y, whose absence remained a relief. Some of the others I remembered from prep week reissued their drinks invitations, including one delegate who had the gag-inducing habit of using female circumcision as a metaphor in his statements. There was also a young Malian, who addressed Marika and me with wandering eyes and a leering "Hello, ladies" it was best to avoid.

I continued to act as bodyguard and security detail for Marika, whose line of work as daily subsistence allowance collector and distributor seemed a hazardous one. When she needed to top up her cash supply, she asked me to accompany her to ATMs. Nothing made me feel conspicuous like carrying around too much cash. I didn't envy Marika the distinction of being known as "the lady with the money." Marika managed it with a kind of determined ease. I sat with her as she counted out South African currency to the LDC delegates whose attendance was supported by the International Institute for Environment and Development.

Wednesday's group included Dr. Y—an interaction I'd been dreading. Seated next to Marika, I watched him intently cross the lobby. My tension rose in direct proportion to his proximity. By the time he was standing directly in front of us, the lid under my right eye was twitching. He avoided looking at me when he sat down. Most of the negotiators we saw handed over their receipts for reimbursement, collected the money Marika counted out, and left with a smile. Dr. Y, of course, was not so pleasant.

"The amount isn't enough," he practically yelled at Marika. "It's half what the others get. Give me more."

He did not return her overt politeness when she explained in a calm, measured voice the criteria set forth by IIED's donors. The UN funded three LDC delegates per country to attend the COP. They were given a lump sum from which they paid for their hotels and any other expenses. Many opted for cheap places, budgeting so they could take the surplus back home. IIED supported a handful of additional LDC negotiators to increase the group's numbers. Marika booked their hotels and gave out cash to cover only daily expenses. The rate was enough for good meals, but it wasn't the flexibility of the UN funding. This I had heard, though Dr. Y's agitation was by far the frankest.

I just hoped that Dr. Y's plan to ignore me lasted through the rest of my time in Durban. He said nothing to me before he left the lobby, and it felt so nice to see him walk away.

After he left, I resumed the primary activity of my twelve-hour working day: assisting Pa Ousman to coordinate and speak on behalf of the LDC Group and to help Achala and Marika as best I could. Mornings began with Pa Ousman in the LDC office, settling his schedule for the day. After that, I went where I was asked. Taking notes, printing things, and buying Cokes—the chair's most requested drink order should he ever have a minute between meetings—occupied most of my time.

"Drinks as requested," I announced, arriving back to the LDC office. "I also bought some muffins and nuts. I noticed that none of you stopped for lunch."

Bubu looked up from what he was reading to laughingly observe to Pa Ousman, "She's not used to people going hungry."

I put the array down on the office's central table before transferring the requested items to the right people.

"You know that's basically poison, right?" I tsked when I handed a Coke to Pa Ousman.

"Poison," he scoffed. "Of all the things to worry about, Coke is not one of them." Pa Ousman laughed as he pulled back the tab.

I moved on to set teas in front of Achala and Bubu. "Speaking of poison," my teasing continued. With a disapproving shake of the head, I counted out four sugar packets for Bubu. The laughter his observation had started ran on.

It was nice to be on joking terms with the Gambians. After all my trepidation over the events of last week, I enjoyed our comfortable rhythm of completing the work of the LDC Group. I also liked having Marika to talk to. As a fellow first-timer, asking questions, making introductions, and exploring were easier with her around, though I did miss seeing people who had known me for longer than ten days.

My fellow Brown University students arrived at the convention center in waves. Some came during the first week of the negotiations; others were scheduled for the second. I told them that my work as an assistant was far from glamorous but the LDC chair and his team were nice people. And Professor Roberts was right; when I was tailing Pa Ousman, all layers of the negotiations were accessible to me. The students watched with large eyes as I emerged from meeting rooms in the wake of special envoys and ambassadors, a PARTY badge around my neck.

My closest grad school confidant, Becca, emailed the week before her flight. I told her not to bother booking a place for her first night. *Stay with me*, I wrote. *I have so much to tell you!*

I rushed to leave the conference center as early as possible on the day she landed. My coming back to the hotel room that evening woke her from sleeping off the flight. She sat up in bed to give me a hug. I gushed, telling her everything: meeting the LDC negotiators, working with the chair and the IIED crew, not knowing what the hell was going on . . . She pulled back her long brown hair to listen. As it all poured out, it was hard to believe that less than two weeks ago, we had been in class together. Like any other Thursday.

"Wow," she said when I finished. "You've been busy."

"It's been something," I said. We sat cross-legged, looking at each other from across the aisle separating the hotel's double beds.

"How are the negotiations going?" she asked, shifting focus. "What's happening?"

"I'm not entirely sure, to be honest." I chuckled.

Shadowing Pa Ousman meant I bounced between topics so frequently that I lost the plot on everything except the general tone. Things were tense. I could read it in every room I trailed him to. What I had to report back about the negotiations' progress was not optimistic.

"Everyone's talking about what to do after the Kyoto Protocol," I started.

"It ends next year, right?"

"Yup—in December."

Two years earlier in Copenhagen, the UN had failed to negotiate a new treaty. While the negotiations last year in Cancún had saved the climate talks from collapse—and established a voluntary forum for nations to continue reducing emissions—no unified structure for cooperative action existed beyond 2012. So expectations for this COP were high, just as Minister Nkoana-Mashabane had said. Durban marked a crucial moment for the climate negotiations.

"They're all asking the same questions," I continued. "Do we create a new treaty? If so, what does it look like? When will it start? Nobody seems to have the answer." In nearly all the meetings Pa Ousman had with Europeans, islanders, Latin Americans, and the COP presidency, the way forward was agenda item number one.

Becca's eyebrows shot up in disbelief. "What? Nobody?"

"Well, no," I clarified. "It's just—nobody has an answer that everyone will agree to."

To date, the negotiations had focused on reducing developed countries' emissions—the likes of the Europeans, the Nordics, the United States and Canada, Australia and New Zealand, Japan. They had polluted first and worst, generating the wealth and ability to make cuts. Americans infamously did not ratify the Kyoto Protocol, and getting the world's largest cumulative emitter to do something was an obvious must-do on the international agenda. Still, how nations should approach a treaty now, nearly fifteen years after defining the Kyoto Protocol, remained unclear.

"I don't know what they're going to do," I worried aloud. "The conference ends next week. How can we just not have an international climate agreement?"

Across the aisle from me, Becca's head tilted in thought, mirroring my own. No one country could tackle climate change alone. The scale of the challenge was just too big. To beat the threat, every nation would need to bring emissions down. And some, like the United States, really should act much more quickly than others. As developing countries rightly indicated, wealth and ability remained unequal. Countries would have to work *together* to enable the most drastic cuts possible and to keep people safe in our warming world. We needed a coordinated response.

"That's crazy. This is crazy," Becca yawned, after a moment.

At a loss, I couldn't think of anything else to say except a sarcastic, "Welcome to Durban!" my hands raised in an exaggerated shrug.

"Great, thanks." Becca yawned again. Then she slumped back down in bed with a resigned "Night."

———————

After a weekend that included a truly epic delegates party—a beach rave organized by the NGO representatives—time started to move quickly. Pa Ousman's twelve-hour days were now fourteen hours full of back-to-back meetings. The days Becca spent in Durban felt like a disturbingly short blip. I wished her trip hadn't been so short and that I wasn't the only one booked to stay a full three weeks. Marika's time at the session wound to a close too. The LDC office felt empty without her, especially as it had grown less occupied with each passing day. Longer and longer hours were taking a toll on people's willingness to congregate and cooperate.

Just before 7:00 PM on the last Tuesday of the negotiations, I set up the room for the LDC Group's coordination meeting and projected a shrinking agenda. Fewer and fewer negotiators had updates to report back. Most of the regular business was concluding. What remained were the tough questions about where the world went from here. When the meeting ended, I joined Bubu, Pa Ousman, and Achala's huddle at the top table.

"The presidency's called the indaba to begin at 8:00 PM," Pa Ousman was saying. To encourage dialogue, the South African presidency had initiated a forum called *indabas* a few days ago. Minister Nkoana-Mashabane explained that an indaba referred to a discussion held by a group of elders, usually sitting under a tree. I pictured a sunlit field arrayed with people in brightly colored clothes nodding concurrence.

"Ay! 8:00 PM?" Bubu yelped reluctantly.

"You go back," Pa Ousman said. "We'll stay."

Two feelings hit me simultaneously—pleasure at being included as a useful member of the team and disappointment at the thought of missing dinner. Apparently, the conclusion of the evening coordination meeting no longer signaled the end of my day. I gathered my things and followed Pa Ousman and Achala back out through the halls of the convention center. At least I would get to see what an indaba looked like.

When I entered the designated room in Pa Ousman's wake, I laughed at how wildly my imaginings differed from reality. Overtired diplomats sat around a square of tables, raising their voices in dead-locked arguments long into the night. For several evenings, I watched the same scene repeat itself around an increasingly thinning delegation until about 2:00 AM. Though the South Africans provided chicken and fries, facilitating effective dialogue had its limits, and with each passing hour, more and more people trickled out the door.

Countries couldn't agree on what the future international effort to confront climate change should look like. Did they need a new treaty? An extension of the old one? Something else altogether? Under it, which countries should do what? Thus far, only developed countries had taken on binding commitments. But today's geopolitical landscape was different from that of 1992, when the original United Nations Framework Convention on Climate Change categorized which developed and developing countries would act. Five years ago, in 2006, China had overtaken the United States as the largest annual emitter of greenhouse gases, and now several other countries the convention classified as "developing" were regularly included in the top emitters list. Defining post-2012 action that looked different pitted those classified as developed countries against developing ones.

And which countries would pay for all of this? The Gambia couldn't even send enough delegates to the talks, let alone afford climate-smart infrastructure. In 1992 developed countries committed to finance collective action under the convention. The most recent iteration of this commitment had come two years ago, when they pledged to mobilize $100 billion per year in climate finance by 2020 and created a new depository to handle these finances. Serious questions remained, though, about where this money would come from and where it would go, as the Green Climate Fund still existed only on paper.

We had reached an impasse only consensus could overcome. Someone had to step forward with a solution and articulate it in a way that everyone would agree to. There was no voting, no "majority rules." Decisions were final when no nation objected, for here "silence means consent." The American and Chinese attempts produced hours of fruitless debate. Sitting behind Pa Ousman, I watched in frustration whenever they took the floor. Power was the ability to inspire wide-ranging agreement rather than bend others to your will. Moral authority could rally strength of numbers, so the vulnerable spoke with influence the wealthy could rarely command.

This dynamic hadn't fully sunk in. I was mostly trying to stay awake, motivated by the dwindling number of remaining COP days. The prospect of the world without a coordinated approach to dealing with climate change was scary. I was just so tired that worrying about it was becoming more and more difficult, as was remaining indefinitely in the indaba. I took breaks whenever possible.

Alone in one of the convention center's eleven restrooms, I told my reflection, "Only one day to go."

I knew all the ladies' rooms now. The one I cried in during prep week was downstairs by the plenaries. I avoided that one. This restroom was upstairs, hard to find, and nearly always empty. I decided to prolong my rare moment outside the indaba and headed to the LDC office. Its door closed with a satisfying click, leaving me alone to tidy up. I recycled the papers left lying around, pushed in the chairs at the central table, and logged people off the two desktop computers. The process was restoring my sense of order and normalcy, almost to where I could imagine a typical Thursday evening at 9:00 PM.

I jumped at a knock on the door, scrambled to reclaim my shoes and a veneer of professionalism. It was John Ashton, the United Kingdom's special climate envoy. Pa Ousman had met with him several times over the past two weeks and I recognized him immediately, remembering his assistant's name. I doubted he knew mine.

"Good evening. Is Pa Ousman available?"

He appeared to be in a hurry, eager to say something.

"I'm afraid he's in a meeting with the COP presidency," I said. Entering, he looked disappointedly around the empty office. "Can I take a message?" I stood waiting and assumed that jotting something down was all I could offer. I was wrong.

John spoke at length about the importance of finding a way forward. Tomorrow was the scheduled conclusion of the negotiations, and there was talk of abandoning the dialogue altogether. My first COP could well be my last COP. *The* last COP.

"Pa Ousman must say something," John went on. "Without the LDC's leadership, this process falls apart."

I nodded, awestruck.

The United Kingdom's climate envoy believed that Durban could end the UN climate change negotiations and that Pa Ousman was key to preventing that. And I was the person he was telling this to. Me, a grad student out of her depth who had never before set foot in the United Nations. A piece of the world's future—whether we would see a new dawn of cooperation—rested, however intangibly, with me and what I did with this information.

Once John departed, I tore at the top page of the notepad in my hand, starting again with the most compelling order and making sure to capture as many details as possible. Then I hurried to locate Pa Ousman, explained John's sentiment, concern, and where to find the British delegation. I watched Pa Ousman gather solutions-oriented delegates around him, opening and closing the LDC office door for them again and again.

The time remaining for countries to decide had run out. Either those huddled in the LDC office would agree to something or no one would.

After talking through what was possible, the LDC Group formally allied with the island states and the European Union to propose the negotiation of a new, legally binding treaty. Under this new treaty, all nations would agree to reduce their greenhouse gas emissions. This would be done on the promise of extending the developed countries' Kyoto Protocol commitments and getting the pledged money into the Green Climate Fund.

Pa Ousman's alliance represented a solution agreeable to the most vulnerable and a wide contingent of the wealthy. What it needed from Durban was the world's commitment to its vision and some agreed milestones toward realizing it. The LDCs, the EU, and the island states were ready to move forward. This was the proposal they brought to the indaba, where they urged others to join them. I saw the South African presidency come alive with hope.

"Colleagues, there is much to consider," said Minister Nkoana-Mashabane. "Let us meet again tomorrow."

I tossed through a restless sleep, glad there was something to consider but too unsure for optimism. Extending the talks meant that when I came in on Friday, I didn't know where to start my countdown. December 9 was supposed to be the last day of negotiations. Flights booked months in advance were abandoned by some and boarded by others for whom rebooking was not possible.

When Bubu couldn't postpone leaving any longer, he clasped my hand. "You'll stay, yes? Help Pa Ousman?"

"Of course," I promised. "I don't fly back until Sunday evening."

Achala reluctantly departed as well while the indaba continued, unceasing. The negotiations became a kind of unending vortex, where people spoke in continuum. Yet their decision—should they ever reach it—meant too much to leave, no matter how exhausted I felt or how tedious the process of getting everyone to agree became. Friday turned into Saturday morning. Morning turned into evening. And eventually Sunday dawned with us still in the convention center.

Sensing consent, the presidency corralled us toward the optimistically named Closing Session when they thought there was enough common understanding to put something down on paper. Pa Ousman and I headed to The Gambia's table. The windowless plenary hall revealed no hint of night or day, just the trickle of people out the doors—finally pulled under by sleep deprivation. Time had ceased to mean much. I couldn't tell if it was 2:00 AM or 2:00 PM, and, at this point, it really didn't matter.

"You can go if you want," Pa Ousman told me. "Get some sleep."

Though I knew my usefulness was limited, I couldn't leave him there alone. Not after we had spent nearly every waking moment together for the past three weeks.

"I'll stay," I said. "Can of poison?" A walk would help. Stretch my legs a bit. Pa Ousman nodded his thanks for the offer and continued to work. When I stood to leave, the live screens flashed the availability of a new draft decision, two weeks and two years in the making. "I'll pick that up too."

On my way back, carrying two copies of the text and a Coke, I noticed movement across the room. A huddle of people was growing around India's seats. Negotiators from China and the United States were pulling up chairs across from their Indian peers, wearing intense expressions. After I handed him the text, Pa Ousman rose to join them, scanning the pages as he went.

I tried to make out the words on mine. It used what I'd come to understand was the UNFCCC's standard layout, only it had the word *Decision* in bold across the top. The paragraphs below were run-on sentences. My sluggish brain had to read through them several times before it could filter out the unnecessary words and decipher meaning.

The UN agreed to negotiate a new treaty on climate change that each of its 195 member countries pledged to contribute to. The treaty would not come into effect until 2020. But, for the first time, the world's biggest greenhouse gas emitters—the United States, China, and India—would be legally obligated to make cuts and protect the vulnerable.

The future climate agreement would apply to all.

I smiled at the words, trying to drink them in, too tired to feel much of anything except relief. The huddle across the room expanded. Though

I couldn't make out the words, I tracked the rise and fall of volume. Minutes passed and the grouping grew until people had to stand on tables and chairs around its periphery to see in.

I watched and waited.

Must be a half hour gone now. Some of the Chinese delegates were holding up their palms. The Europeans were on their feet and then the Americans too. The huddle finally began to disperse, moving back out from India's table. Pa Ousman drifted back to The Gambia's seats.

"So?" I asked.

"We have agreed," he nodded. "We will negotiate the new treaty by 2015."

Using the last of my energy, I eked out an eye-crinkling smile.

"The UN has agreed," I repeated.

Thirty-six consecutive hours after the talks' scheduled conclusion, an exhausted Minister Nkoana-Mashabane took the top table and looked out over the near-empty room. I bet she prayed no delegation would call for a quorum check. The chances of proving that half the world was represented were slim at best. Minister Nkoana-Mashabane gaveled the adoption of what she called a landmark on our comprehensive outcome and brought the COP to an end.

Delirious with sleep deprivation, I collected my things and followed Pa Ousman out of the convention center one last time. We exchanged no goodbyes, just slowly drifted apart. I didn't question my overwhelming sense that we would meet again.

4

BREATHLESS

Kelso, Washington
December 2011

I WOKE AS THE PLANE dipped below the familiar cloud barrier covering Portland, Oregon. Muddled, I blinked back the journey: South Africa to the Pacific Northwest by way of Rhode Island was a long trip. I took in the trees that lined the wide arch of the Columbia River, the wisps of mist rising from the water caught in the hills aglitter with houselights and the sunset reddening Mount Hood's silhouette along the eastern horizon.

Outside of the arrivals door, I looked for the blond head and waving arms of the woman who remained my best friend. I had intended not to return to Kelso after high school, but Seattle was an easy distance for Noelle and me to close. Frequent visits shaped my years at undergrad and the one after, when I stayed for AmeriCorps. Laughter over memories old and new followed our exploration of the Pacific Northwest, no longer that of children but of adults who could go where they pleased. My farthest relocation to date, the cross-country move to Brown, had stretched our habits. Half a year had passed since I had moved east, an eternity when measured in time lost encouraging each other through the latest exercise craze or comparing reactions to the newest health food diet. Though my immediate family remained in Kelso too, it was seeing Noelle that I looked forward to with unburdened excitement,

65

time with her that I most wanted and most missed. And Noelle stood, waiting expectantly, outside Portland International Airport.

Our coat collars and the frenzy of the loading and unloading zone muffled the "Noelle!" I shouted in her direction and the "Welcome home!" she yelled in mine.

Her husband, Kevin, stowed my bag while we hugged it out. Noelle had married Kevin six months prior on a pier overlooking the Oregon coast. I liked them together. I had from the first time I met Kevin, years ago in the apartment Noelle moved into after high school when she worked as an assistant in a lawyer's office while finishing her degree. On one of my trips down from Seattle, she asked her new boyfriend to meet her best friend.

He needed to make a good impression.

Not one for small talk, especially when meeting someone for the first time, I imagined Kevin found the prospect of getting to know me daunting. Noelle and I could easily do nothing but talk. But he knew our meeting was important because to Noelle, I was important, so much so that when I first shook hands with the muscular, six-foot-two apprentice carpenter standing in Noelle's living room, he looked, above all else, nervous. I had liked them together ever since. Kevin now drove us home to Kelso along the dark highway, windshield wipers pushing away the December rain.

"So, what's new and exciting?" I asked, stretching out across the back seats.

"Have you heard that Country Village's café is closing?" Noelle said at the same time Kevin shrugged and said, "We're getting a Planet Fitness." I saw their heads swivel toward each other.

"We are? When did you hear that? Why didn't you tell me?" Noelle turned.

"Probably because you were yammering on about something or other and I couldn't get a word in," Kevin replied.

They quipped for a few miles while I laughed comfortably, realizing how much I had missed them. Our drive ended one block past the weeping willow that still shaded Noelle's childhood home. Rather than pull right into her former driveway, Kevin took the next left onto a narrow road I had never explored. The lane ended at their corner lot townhouse.

It consisted of a two-storied collection of slanting roofs, wood siding, wide windows, and their new lawn that enjoyed the protection of a large cedar tree. A wreath Noelle had made hung on the door.

I instructed them both before they married that any future properties would need to have room for me. Noelle had laughed, unquestioning. Kevin had half smiled, perhaps only just realizing what he was getting himself into. He eventually nodded in agreement. True to their word, Noelle gave me a tour of their new place that ended in an upstairs office complete with an air mattress.

"Your room," she waved.

"For me?!" I beamed in mock surprise. When Noelle fussed, I thanked her sincerely. "It's perfect; I love it."

I spent most of the holiday sleeping off the jet lag and long nights of Durban, which felt more like dream than reality with each passing day. Waking at odd hours, I dismissed the visions of negotiating climate change with African delegates as fantasy, only to remember in a rush and smile—astonished that such remembering was my own. I had really been there, negotiating a new climate change treaty. Wild.

Days in, Mom called to ask when I would come home, bringing me back down to earth, reminding me that unending slumber parties with my bestie weren't all that waited for me here. I knew that Mom missed me and that my vagueness about where exactly in Kelso I intended to stay hurt her. But how could she not understand why I kept my distance?

I sighed, defeated, knowing I would go eventually.

I could hear the reason in her voice.

———————

Life in my parents' house was changing. Christmas morning saw the same faces gathered around the tree, retrieving presents from the same stockings that hung every holiday season. Chanteal was there, just as she had always been when I returned for Christmas. These days, she lived at home, commuting to campus in Vancouver. An inch shorter than I, she wore glasses and kept her hair tied back in a low bun. When the snow began to fall, sticking to the trees and blanketing the world outside in quiet layers of white, she was up lighting the fireplace.

Mom still worked at the local hospital and still wore an apron to make buckeye candies. She enjoyed my presence so much that when Chanteal, Mom, and I played games, she let me cheat. People who cannot spell should not win at Scrabble so frequently. Afterward, we still went out to walk the road.

The change was Dad. He was ill.

In my early teens, Dad got sick and spent a few weeks in the hospital. At first, the doctors thought it was a bad case of bronchitis. Only it didn't go away. He dropped weight and coughed blood. And even after he took regularly administered medication, things didn't turn around. The doctors couldn't figure him out. They diagnosed his illness as another type of persistent lung virus, which perhaps he picked up when he traveled for work after his year in Virginia. His condition eventually stabilized and, armed with prescription steroids for the lung inflammation, he came home.

It had been a decade since, and the lung virus remained. Doctors now described his condition as a type of chronic pneumonia: bronchiolitis obliterans organizing pneumonia, commonly called "BOOP." Day after day, his lung tissue succumbed. Medications managed the symptoms, but nothing—save new lungs—could cure it. He wheezed. Then he coughed. At first, infrequently. Then, so constantly that he needed to sleep with an oxygen tank. The athletic part of his identity slipped away. The control he loved so much slipped away. He needed air constantly now. He couldn't work.

Dad's reliance made him meaner, even more difficult to live with than I remembered. The glimpses I had of home life in the years after I moved out were traumatic, my short stints down from Seattle throwing the nightmare into starker relief. I generally thought myself good at reading people. I was not good at reading him. I never knew what he would do or how he would react, what would set him off. This combined badly with my propensity to ask impertinent questions and high tolerance for risk. And I wasn't the only one. The circle of people Dad spoke to contracted every year. He had stopped talking to Granddad, Uncle Charles and his family, and most of our neighbors. Asking him why was a fool's errand.

But he was the kind of sick Mom wouldn't walk away from. Without her and her job and her health insurance, he would not survive. I didn't understand this either. I had always struggled to really comprehend Mom's thinking. Her opinions were often dependent on who was asking for them, a useful, if alienating, skill for living with a man like Dad, one my forthrightness would never allow. Mostly, though, she retreated. Perhaps she felt safer within the confines of her mind. I hoped things were happier in there, and I wished saying no to coming back was easier, that seeing her didn't involve such difficult circumstances. The truth was, I missed her too.

Come Christmas, Mom, Chanteal, and I went to the neighborhood party. We sung the carols and unwrapped the white elephant gifts in celebration of the season while answering polite questions about Dad's health. We went to church—where Mom's head bowed long in prayer—without him too. Christianity had structured my education until age twelve. I went to Bible camp most childhood summers and could still recite the sixty-six books in order, as I could the Apostles' Creed. At thirteen, I had been confirmed a Lutheran in the very church where Mom now prayed.

I had always believed in God.

And He was everywhere.

From the daily Pledge of Allegiance we recited in public school to the teachings of Dr. Martin Luther King Jr., faith for me was ubiquitous. During my rural upbringing, it went nearly unquestioned, Christianity the prevailing opinion of how the universe worked.

In the pews, I shared the peace by shaking hands with families I had known my entire life, easily answered questions about what I was doing these days, asked after their lives and properties, and echoed the answers Mom gave about Dad.

"How's Glen doing? We haven't seen him in ages."

"He's hanging in there," Mom said. "Thanks for your prayers."

Then we drove home in silence, prepared to say nothing more about it.

Dad's illness fueled a relentless anger. My parents' house teetered around his mood, strained quiet the calmest atmosphere possible. His eyes bulged, and the smallest things provoked him. Dirty dishes were stacked along the back edge of the kitchen counter until one of us washed them. Once cleaned, his clothes were to be folded, delivered to his room, and placed at the foot of his bed. Dinner was to be served at 6:30 PM. He would punish any breach of this protocol.

The topics my father and I could safely discuss were always limited at best. Our conversations now held a maximum word count of fewer than ten.

"Why can't you follow simple instructions!"

Man alive, I couldn't even get through a morning's bathroom routine without hearing Dad yell. The noise was coming from down the hall. Once I finished, opening the bathroom door gave me a clear shot of him. He and Chanteal were at the end of the hallway, standing by the laundry room doors. I saw the problem when I got close. His clean clothes were neatly folded on top of the dryer, not at the foot of his bed. The living room door swung open, and Mom's footsteps drew closer to the growing volume.

"You're worthless!" he shouted at Chanteal.

I had lived away from his house and his rules for six years now. Six years of life outside Kelso, Washington. A great, wide world of places and peoples and lifestyles so different from mine that they moved me to question, inspired me to revel in the freedom of drawing my own conclusions. Six years of figuring out who I was and who I was not.

I corrected him without hesitation. "You're not worthless."

"What did you say?"

I recognized the threat, so often repeated throughout my childhood, and the effect it was meant to convey—fear and submission. But I wasn't a child anymore. For starters, I looked down at Dad now. The inch I grew in undergrad combined with the height he lost to illness made me the tallest member of my immediate family. Given how thin he had become, I likely outweighed him too. Where I used to see indomitable strength, I now saw instability and a rage that defied even his own logic. Instead of shutting up or backing down, as he meant me to, I just kept speaking.

"I'm talking to my sister," I said, surprising even myself with the confidence of my tone.

He came at me, forgetting what the effort would cost him. The oxygen pack he wore attempted to lurch with him, the tube that in a split ran into his nose swung and strained. I heard the coughing start, the restricted pulls of his lungs, the hacking desperation for oxygen. He backed away, eyes filled with indignation, still eager to exact his anger on those he deemed more vulnerable than him.

Carless, I packed my bags and asked Chanteal for a ride. We didn't talk about it. The postconflict expectation to say nothing and move on remained. It may have made coping easier in the moment. Yet the determined effort to avoid discussing anything unpleasant atrophied our ability to connect and bred powerlessness and isolation. In that house, I felt mostly alone even when occupying the same room as my immediate relations. Even with Chanteal, what little there was between us seemed to exist only when we were forced together, the loneliness exacerbated by the expectation of closeness.

My family lived adjacent, rather than intertwining, lives.

Chanteal dropped me at Noelle's before continuing to town in the frost of the December afternoon. Noelle, Kevin, and I made dinner in comfortable familiarity. Afterward the ladies took over the living room with *Gilmore Girls* reruns and nail polish. Kevin went upstairs to play video games. I changed into sweats and balled up on their couch.

"He reminds me of Darth Vader," I told Noelle apropos of nothing.

She looked up from her toenails, raised an eyebrow and waited. "Your dad?" she eventually prompted.

I nodded.

"*Shhhhhoopah. Shhhhhoopah,*" I breathed heavily in and out, exaggerating the noise of his oxygen pack. Did only Skywalkers get light sabers to fight their father who couldn't breathe? "At least it's easy to hear him coming."

Noelle laughed out of pity. I needed the humor to deal, and she let me have it.

Noelle didn't ask me to make sense of my truly unfunny attempts. Even though the pretending of it all drove me crazy, I carried the strained silence that dominated my parents' house into my other relationships, telling distressful things only when they were too fresh to hide or when veiled, however poorly, in humor. The anger and betrayal, the fear and

pain, the unconquerable depression that ruled much of my childhood—
they all went unspoken, unsuccessfully buried within myself. After ten
years of seeing the crests and falls of my relationships with my family,
perhaps Noelle knew I wasn't going to talk about it.

Or that I simply didn't know how.

———————

Christmas done, I took the train up to Seattle, happy to put some distance
between myself and my father. Besides, I missed the city and the traditions
I had missed that year, like the pilgrimage to Jan and Maynard's place for
Thanksgiving, my favorite holiday. Maynard was an old army buddy of
Granddad's. He and his wife lived in a brick house with a leafy backyard.
Every year, they invited a small crowd over for an amazing spread. Since
we were little, Chanteal and I had walked their golden retrievers in the
park during the long hours between our family's arrival and when the
food was ready. Stan, another former army buddy, carved the turkey, and
his son ate more than everyone else before surrendering during the pie
round. The adults moved on to playing pinochle, while the younger set
took the best-pie argument into the living room where *The Wizard of Oz*
was playing. For the record, blueberry pie was always best. Sweet potato
had nothing on blueberry.

I would miss the demonstrations remembering the work of
Dr. Martin Luther King Jr. this year too. The words to the Black national
anthem and "We Shall Overcome" were inscribed on my subconscious
in early childhood. I could recite them without conscious effort. As kids
we learned to shift weight during long speeches, to march and crowd
magnify, to call and respond. If the holiday found us in Seattle, we
convened at Granddad's house for pound cake and stories of hearing "I
have a dream" ring out across the National Mall. I soaked in the understanding that the legacy of nonviolent, civil disobedience was part of our
inheritance. That demonstrating and personal freedom—particularly the
right to vote—were inalterably tied together and must never be taken
for granted, that this ongoing struggle encompassed my American life.

At King Street Station, I hurried through the Seattle bustle. I was late
for a reunion of sorts. The streetlights came on and the rain started just

as I arrived at the correct multistory building. Two voices echoed through the buzzer's intercom. I followed Michelle and Erina upstairs, waiting to officially begin our time together with what would happen next.

Fiddling with her iPhone speakers, Michelle turned up a beat. Michelle was witty and full of opinions. While we all more than qualified as tremendous nerds, in undergrad Michelle was a neuroscience nerd, which made her overachievement even more impressive. She had an amazing singing voice and lovely thick hair that she kept glossy and short. Her impersonation of her Sri Lankan parents never failed to make me cry with laughter.

I didn't clock which song began before I started jumping in rhythm. Erina rotated out from the central space, moving her arms in time. Erina had a great love of words. Her Japanese parents sent her to language school as a kid, and she was learning French now. She baked as a form of meditation and was in all things infinitely practical, except perhaps in her obsession with Hello Kitty. Neither she nor Michelle topped the five-foot-five mark, but they were not ones to be easily looked over.

We had met in Rome four years ago, and our differences were among the things that inspired our friendship. Proprietors always knew we were American—even before giving ourselves away with our terrible *italiano*—for nowhere else would convene such a diverse student group. In many conversations over truffle ravioli, Michelle, Erina, and I puzzled over the country that both brought our families together and made racism part of our daily lives. We made up racist hick names for each other after an evening of story swapping. Our varying perspectives helped us turn what was incredibly sad into something riotously funny. Shaniqua Lou, Hindi Sue, and Sushi Anne roamed the streets of Rome together for the remainder of that summer. We had been close ever since.

As was our tradition, we danced in a shimmying, rotating circle until we noticed exasperated glances from Michelle's neighbors across the alley. When their blinds went down, we lost it, laughing until we collapsed into the furniture. I soared in the warmth of our friendship, the trust that after years and across continents we would find each other again and again. I forgot about my father and the stress of graduate school. I was Shaniqua Lou and I was with my chosen family under the home of our overcast sky. And that was all that mattered. My world may have begun in violence and fear, but there were other ways to live, different from how life began: Loud, ritual dances. Friendships where I belonged. People who made the crises manageable, even if they didn't know that was what they were doing.

I stayed a week in the apartment Michelle shared with her sister, Andrea. Erina came and went from her parents' house north of the University District. We laughed at old times and filled each other in on our changing lives. Erina's stories were about life in France, where she worked as an English teacher. Michelle detailed the manic pace of the University of Washington's medical school. And I described my first semester at Brown, including the strange trip I took to South Africa.

We yelled, "Happy 2012!" when midnight struck on New Year's. And when Erina's and my departures drew close, we brainstormed where we would next meet. Erina's reminders that we were welcome to Paris sent eyes rolling in her direction. "I wish," I griped, thinking with despair about all the borrowed money that passed straight through my bank account to Brown's to cover my tuition. After one last group hug, Erina headed for the airport, and I caught the train back to Kelso where Noelle would pick me up and the wrench of saying goodbye would start all over again.

The brick quads and historic roads of Providence, Rhode Island, welcomed me back to another semester of grad school with a bracing chill. It struck me again how different I found the place, as I made my way to Brown University.

When I had moved here in summer, the flatness of the landscape enveloped me. I couldn't get my bearings without mountains for orientation. In my opinion, none of Providence's seven hills were high enough to afford a sufficient view. The land just stretched on uninterrupted into the green inland or out to sea. The place itself brimmed with history to such an extent that placards marked the sidewalks. Cemetery tombstones crumbled with age and steeples were often the highest points in the landscape. I remembered hearing crickets sing those first evenings against the sky and thinking that the night felt different here.

When autumn arrived, I drank in the midsixties sunshine while marveling at the colors of the leaves. In the Northeast, people filled huge brown paper bags with them once they fell. I reviewed how scientists measured the concentration of carbon dioxide in the atmosphere. A lab in Mauna Loa, Hawaii, recorded the parts per million among the contamination-free vastness of the Pacific Ocean. And even there the concentration peaked and troughed in a yearly rhythm—reflecting the global exhale of carbon caused by the leaves falling from the forests of the Northern Hemisphere. The million embers of painted trees were not observed in the Pacific Northwest. Coming from conifers, I could never envision the scale that made atmospheric measurements resemble a respiratory chart. Then I witnessed the mainland forests cease to infuse carbon into their limbs. They shed leaves on a continental scale, leaving our overrun climate to struggle for breath. I passed the bags of leaves stacked neatly on the sidewalks and wondered what happened to them.

What would happen to us all.

Thunder rolled in the coming winter, a season that turned on porch lights. Students donned the longest, puffiest coats I had ever seen. I had yet to experience what made them necessary. I thought perhaps the winters people warned me about when I said I was moving to Rhode Island wouldn't materialize. Then the temperature started to fall. A violent, bone-invading chill shook me so hard my skull rattled. I learned expressions like *nor'easter* and that snow could remain unmelted so long it looked dirty. As someone who preferred weather that was too warm rather than too cold, I honestly wondered how people survived here before central heating.

When I returned in January, Providence felt like a frozen hellscape.

In a well-intentioned effort to lose some holiday weight, fellow first-year Becca and I devised a running schedule. Half the days we planned to meet at the high school track, I bailed out. Negative twelve degrees Fahrenheit invaded my teeth with a painful frigidity. Such temperatures rendered all function impossible. I put on every layer I owned that was fit for running, only to get half a block before shrieking and sprinting back inside. "I can't do it, Becca," I said over the phone.

"What?!" I knew that she, a Massachusetts native, thought me completely ridiculous. "It's not even that cold. The sun's out," she said disbelievingly.

"Look, any amount of rain is fine. But sun *and* ice is just a cruel mind trick." My feet, even in summer, were rarely warm to the touch. I thought now they might actually fall off. Her laughter came through from the other end.

My schedule of methodology courses overlapped with hers in most instances. On the docket for this semester were environmental economics, statistics, and the geographic information systems course. I missed the visuals of studying architecture and thought GIS mapping would help break up the numbers focus of the other two. There was also the weekly climate and development lab Professor Roberts convened, where I had raised my hand months ago.

The daily thinking I did about climate change felt, in some ways, wholly unconnected to the world I glimpsed in Durban. I had hoped that my role at the UN would continue or lead to something else. But I couldn't see a way back in—even as I considered thesis topics focused on examining the outcomes of the climate negotiations. Even though I'd held one in my hands, UNFCCC decisions were back to being things I read about, not things I saw made, let alone had a hand in.

———

Then, on an unremarkable day in early spring, I opened my inbox to a new email from Achala. She wrote that her role supporting Pa Ousman's chairmanship of the LDC Group was to continue and that she had enlisted Marika to provide coordination support full-time. If I wished to join, she would welcome my help as a research assistant for the next round of climate negotiations.

I blinked at the screen, reading and rereading her message.

It was everything I'd hoped for and yet completely unexpected. Apart from some follow-up emails once the conference officially ended, I had heard little from the people I'd met in South Africa. Achala's abrupt departure from Durban meant we hadn't talked about the future. I was thrilled that she wanted me back. And I assumed her invitation meant that the Gambians were happy with my work too. Otherwise they surely wouldn't have asked me to continue. My time at the UN could remain a singular experience or be just my first at the negotiations. The choice was mine.

I thought about it, that first terrible week and the sleepless stretch of the last. Feeling completely out of my depth. The ridiculously long hours. And the unbelievable position working with the Gambians placed me in. In Durban I had watched in real time the UN agree to its next chapter of international climate action. Followed Pa Ousman right through to the rooms in which governments forged solutions to the crisis. And I could do it again.

I could help stop climate change.

I typed out my response accepting her offer, then pared back the thank-yous and exclamation marks to as professional a limit as I could manage. I happily exchanged emails accepting a place on the Gambian delegation for the foreseeable future. Before I ended the chain, I threw in a casual *Where are the next negotiations? And when do they start?*

Copied in, Marika replied with logistical details, telling me which hotel she and Achala planned to stay in and how to get to the negotiating venue. She added that this was all secondhand information, as she hadn't been to Bonn before either. I learned that while the COPs were the annual finale—the showstopper that garnered as much attention as the UN climate talks ever attracted—the negotiations that took place away from the movable feast were held in Germany. Google told me that Bonn had been the capital of West Germany, the birthplace of Ludwig van Beethoven, and that the price of accommodation was more than I could afford.

I was broke. Again.

In my excitement to continue in the negotiations, I hadn't thought through the financial implications before responding. Unlike COP, where Brown sent and paid for a group of students to attend, the university didn't send people to Bonn. No school budget waited to cover my expenses. And what Achala was extending was an opportunity, not a paycheck. Sure, the continued experience could lead to a paying position someday. But that day was not today, and I was living a borrowed lifestyle.

The amount of debt I watched accumulate in exchange for a graduate education made me dizzy. I worked while studying and had earned a summer study grant from Brown. Even so, there wasn't money left for taking unpaid work-experience trips to Europe. Traveling to and staying in Germany without help was beyond me. It seemed especially foolhardy when remaining in Providence would mean money in the bank and carried no chance of me sleeping on the streets in Deutschland.

The potential gleamed so enticingly though just out of reach. Maybe I could find a nice host family willing to house me on the cheap. Surely it would be easier to find something once I was actually over there. Perhaps record-breaking summer temperatures would mean everyone slept outside in Bonn. The worn circle of risk and reward picked up in my head again, until I confessed to Becca, "I don't think I can go."

She and I were in one of Providence's numerous coffee shops attempting our environmental economics homework. She had asked for an update about Achala's offer, which forced me to fess up about the state of my dilemma.

"Why don't you ask to share Marika's room? You said she was nice to you in South Africa."

I hadn't thought of that. Marika had been friendly, but I was fairly confident even the nicest of people wouldn't offer to share their hotel room with a broke grad student for a multiweek work trip. Marika had a job, after all. She didn't have to slum it like me.

"I don't know. Is that weird?"

"I stayed with you in Durban," Becca shrugged. "And if it's share or don't go, it's worth a shot."

5

FALSE START

Bonn, Germany
May 2012

A FEW MONTHS LATER, I pulled my suitcase away from Bonn's airport bus stop, the address of Marika's hotel in hand. Her response to my groveling email had been all kindness; she'd agreed to let me share her room for the next two weeks. All I had to do was find it. And thank her—massively.

The flight of pigeons above cobbled squares called to mind a nostalgic remembrance of life in Europe. I took in the closeness of the outdoor food markets and the people chatting under café awnings; stared up at the ancient cathedral arches hurtling toward the sky, their bells ringing in the hour. I smiled at the stands selling gelato; remembered what it was like to not know what anyone was saying. Though when I asked, "*Sprechen Sie* English?" most people responded "No," then followed up with English better than my own.

On Pa Ousman's recommendation, Marika booked a place that turned out great for its location and severely lacking in everything else. I laughed out loud when I opened the door to our room. Even with the tiny twin beds mounted to the longitudinal walls, they were divided by a gap of only about a foot. This created the real possibility that I might roll onto Marika's bed in the middle of the night.

Our newness to Bonn gave Marika and me plenty to talk about as we set out to explore together. The city was the epitome of green sub-urbanism, tranquilly located on the banks of the Rhine River. A close winding of streets surrounded a fading, pink town hall whose square attracted a lively market most mornings and whose restaurants and ice cream shops filled the evenings with the noise of happy diners. Radiating out from the city center were a series of other squares adjacent to Romanesque cathedrals, the lawns of the University of Bonn; the central train station, or Hauptbahnhof; and Bonn's claim to fame, the birthplace of Beethoven.

One of the perks about hanging out with Marika was our shared insistence on vegetarian options. She had adopted vegetarianism even earlier than I, childhood Marika intent on playing no part in harming animals when simply eating other things could avoid it. We took to ordering meat-free meals wherever we could find them and discovering new hilarious differences between British and American English. The hilarity made me realize how much I genuinely liked Marika. Continuing in the negotiations would mean working and rooming with her. I counted myself very fortunate that doing so was fun.

Five metro stops southeast of the Hauptbahnhof, the UN's climate change headquarters occupied one of Bonn's two skyscrapers. The negotiations were held around it. Remembering the last few days of Durban, I arrived expecting hubbub—traffic, cameras, lines of people at security, press. What I found was a quiet conference center, one I would have walked right past without Marika's careful directions.

Even though today was the day negotiations were set to begin, I followed Achala and Marika through sparsely populated corridors to the allotted room for the LDC Group's first 1:00 PM coordination meeting.

The LDC chair, Pa Ousman, and his ever-present vice chair, Bubu, stood at the front of a room oddly similar to those in Durban. It held the same long white tables arranged classroom style facing an elevated top and the same tabletop microphones and flags. Pa Ousman and Bubu looked better rested than when I'd last laid eyes on them in South Africa six months ago. Both stood and smiled as we entered.

"Hello, Brianna. How are you?" Pa Ousman asked, shaking my hand.

In contrast, Bubu let out an "Ah! She's back!" and upstretched his hands like I was a long-lost relative. I laughed remembering what it was like to be around the pair then thanked them for the opportunity to belong to the Gambian delegation again.

After we said our hellos, I looked around, trying to remember which names belonged to the faces that, in the long hours of Durban, had grown so familiar. Catching on, Bubu took my hand and led me on a tour of boisterous introductions during the lull before the meeting began.

First stop was Ian, a man with gray hair and glasses. His PARTY badge hung from a string of seashells rather than the UNFCCC's standard blue lanyard. I wondered if they were from the Pacific Island nation he represented.

"This is the professor, Ian of Tuvalu," Bubu reminded me while we shook hands.

"I've been around even longer than Bubu here," Ian said in an Australian accent.

"And Sandra, my sister, how are you?" Bubu asked a nearby delegate. The woman from Togo kissed my cheeks. She kept her straightened hair short and brushed back and usually wore a stern expression. I remembered her as one of the handful of women actively involved in the LDC Group. I also remembered that I had to look up where Togo was, next to Ghana on West Africa's coastline.

"Have you met Manjeet?"

A guy with dark hair, dark eyes, and dimpled cheeks shook my hand. I smiled reading his PARTY badge: Nepal. Finally, a country I had actually been to! He seemed younger than most of the other delegates and less formal. Today he was wearing a polo shirt and khakis. I often saw Sandra and Manjeet together. I knew they were coordinators for the LDC Group and worked closely with Pa Ousman. Now I learned that they were supported by the same NGO, similar to how Pa Ousman was supported by IIED.

The string of faces shaking hands and kissing cheeks continued as Bubu led me around the room. Most smiled in response to the joviality Bubu carried with him. I was eager to get up to speed with the negotiations, not knowing what had happened since we last parted ways. But getting to know more members of the LDC Group was important too. Though I was happy to be an honorary Gambian again, I still felt like an outsider. I made an effort to remember the names Bubu rattled off so quickly.

"This is Mbaye from The Gambia—no, Senegal!" he gaffed, grasping a thin man's arm. I recognized him. I remembered liking the way he used an enthusiastic, "Exactly!" to punctuate his sentences. Mbaye was smartly dressed and wore square glasses. He had long expressive fingers like Pa Ousman, but the rings he wore on them were a different style.

"Bubu!" Mbaye gripped Bubu back energetically. After they had finished, he extended a hand in my direction.

"My sister," he smiled. "Welcome! I can tell by your hair that you're one of us."

Sure, I'll take it. "Thanks for the welcome," I responded.

"Bubu will show you around. How are things back home, brother?" he asked. I remembered Senegal as the country that bordered The Gambia on all sides except for the Atlantic Ocean.

"Fine. Fine," Bubu nodded. "The renovations on the house are still going. My wife . . ."

"Which one?" Mbaye interrupted.

"Number two . . ." he continued without pause.

My eyes widened and I turned my head to look pointedly at Bubu. Both men laughed at my response, bringing the nipping anxiety induced by meeting new delegates to the forefront of my mind. While milling around the room, I noticed several of the men who had asked me out in Durban. I thought it unfortunate that they were negotiation regulars as well. Least favorites included Dr. Y—whom I continued to avoid as vehemently as he avoided me—the delegate with the crushing handshake, and the leering "Hello, ladies" Malian.

Following our run-ins in Durban, Marika had turned an Excel spreadsheet into a sexual harassment tracker. Marika loved a spreadsheet. She now kept a thorough log of every unwelcome invitation, disturbing advance, and unwanted touch. The tracker made a handy list of things to be aware of for any new female colleagues we met. Knowing that each incident was carefully recorded and shared made them easier for me to deal with.

From the top table, Pa Ousman looked over his glasses to call the room to order and greet the LDC Group. I listened intently to the updates, needing to understand where things stood. The UN's primary focus was to interpret last year's decision and lay this out into a program of work. That sounded simple enough to me. In Durban the world agreed to negotiate a new, legally binding treaty under which all nations would act to combat climate change. We had until 2015 to negotiate the treaty, and countries had until 2020 to start implementing their commitments. Surely, we could just start work on drafting it at the beginning.

Wherever that was.

After coordination I followed Pa Ousman and Achala through to plenary—tickled to see the name flags of the whole world spread alphabetically across a hall again and to relive the sense of Olympics-like world peace. I was determined to make a better effort to understand the negotiations this time around. Only once we entered did I realize that they hadn't started yet.

Onstage, a projected agenda read that this was in fact another coordination meeting for the Group of 77 and China. A man I assumed was the chair sat illuminated at a long, linen-covered table. The hall was so large that I couldn't conceptualize this as a meeting. I wondered how anyone was following the dialogue that speakers broadcast around us. Understanding the groups, who they were, and whom they represented felt like a good place to start improving my knowledge of how things worked.

Seated in The Gambia's four designated chairs behind Pa Ousman and Achala, I opened my laptop and typed "G77 member countries" into Google for an answer. Misleadingly, the Group of 77 and China was made up of 134 countries. Together they accounted for 80 percent of the globe's population. Scanning the list of members told me that the man onstage spoke on behalf of essentially all developing countries.

I remembered working with the island states in Durban, opening the LDC office door for dreadlocked ambassadors come to talk to Pa Ousman in the COP's final hours. I typed *small island developing states* into Google next. Their membership was defined by United Nations criteria, like the LDCs. In fact, several LDCs were also classified as Small Island Developing States. The Alliance of Small Island States, a negotiating group, also came up in the search. Their lists of member countries were basically the same. I scribbled *AOSIS Group* next to small islands in my notebook and *all developing countries* next to G77. I tapped my pen against the page in thought. A single country could be a member of all four of these groups, and all these countries were probably represented here in this room.

I tuned in to the conversation around me, trying to connect it to the handful of bullet points on the agenda. I had to concentrate to discern the speakers' meaning from their eloquence. Representatives from different countries were arguing about what they should include in a speech. More specifically, they were arguing about what the G77 chair should say at the plenary due to start in a few minutes, which was why the G77 chair was struggling to wrap things up. Understanding that they were arguing dampened the world peace feeling I'd had on entry. It was the same during the Olympics whenever the athletes literally started beating each other for sport.

G77 concluded and the earth's other countries filed in. Just as in Durban, South Africa's Minister Nkoana-Mashabane took the podium when it came time to discuss the landmark decision she had gaveled through five months ago. Her ringing call for nations to engage constructively brought a surge of déjà vu.

"We must create a credible plan, prioritizing work in such a manner that we finish by 2015," she said.

My eyes shifted to scan the dignitaries at the top table. I happily recognized one other face. The executive secretary of the UNFCCC, Christiana Figueres, sat looking out optimistically over the crowd. From Costa Rica, she was the woman responsible for the UNFCCC secretariat.

Hers was the constant face of the climate negotiations, the one unchanging as they were handed over from country to country.

Group statements began, and Pa Ousman was among the first to echo Minister Nkoana-Mashabane's request. He was eager to start work on the new climate treaty; the lives of his people depended on it. On behalf of the LDCs, he suggested identifying deliverables for each COP, in a three-year work program that would ensure completion. Other speakers followed with their suggestions, most of which appeared similarly sensible.

An hour in, someone started speaking French. I mimicked Achala and reached for the headphones under my seat that offered simultaneous translation into English. Curious, I scrolled through the other channels. French and Spanish I recognized easily. Then Russian, I guessed. Mandarin. Arabic?

Up until this point, everything had happened in English. The signs, the documents, everything. As a native speaker, I hadn't noticed. It struck me that when the world came together, mine was the language of negotiation. I went through an entire UN climate negotiation and it never occurred to me that so many must find the UN climate talks completely impenetrable.

When the UNFCCC secretariat drafted an agenda to move discussions forward, I felt the room tense up fast. Talking through a work plan was one thing. Adopting these words as written was another. The agenda was thought workable by some but not by others. Every group interpreted the Durban decision slightly differently and placed their emphasis on different areas.

At Brown we referred to climate change as a wicked problem, which was not, to my amusement, just East Coast slang for a complex issue. Climate change combined the interconnected problems of sustainability and pollution with many actors, long timescales, great economic burden, and uncertainty. At school we compartmentalized the challenge into three broad thematic areas: *Mitigation* covered the reduction of warmth-inducing greenhouse gas emissions. *Adaptation* meant adjusting to their impacts. And the *means of implementation* were how either was done—primarily with money but also by building capacity and transferring technology.

Reducing emissions was no longer the only UN agenda item of importance. Twenty years on from the signing of the convention, adapting had risen to primary importance for many countries, particularly those in the developing world, which were hit hardest by climate change's effects. Questions regarding how to adapt were closely followed by concerns over where the finances, technology, and capacity to do so would come from.

The group statements continued with some pointing out that even though the treaty did not start until 2020, countries needed to act well before then, and this should be reflected in the treaty proper. When they finished, others argued that the agenda was too focused on reducing emissions. They couldn't agree. And without adopting an agenda, they couldn't start work.

Hours turned into days, and I honestly marveled at the amount of time people could spend deliberating *an agenda*.

Fortunately I had plenty to distract me. As part of her new role at the negotiations, Marika took over managing Pa Ousman's schedule in addition to handling the finances. This left me with more time to edit the documents and statements sent to me by Achala and other members of the LDC Group. After Bubu facilitated introductions, negotiators would wave me down in meetings or when they spotted me in the LDC office.

"Please would you check?" they asked, passing me a statement or speech they had written. I had never valued speaking English as my first and only language as a skill. Here, negotiators looked on with relief in their eyes, grateful to have a native speaker around to look things over. Surely I would know how to say it. I prayed that I didn't screw up their meaning too badly.

With each interaction, I realized just how unequal the negotiations were. Apart from happening in English, LDC delegates were likely to have several UN forums to cover while their developed-country counterparts worked climate change full-time. The unequal body count remained too. Pa Ousman needed people like Achala and her team to support the overworked and outnumbered LDC Group.

I passed the hours in the seemingly endless agenda discussion typing from the back row of The Gambia's square of four chairs in plenary, Pa Ousman and Achala at the table ahead of me. It felt nice to start wrapping my brain around the sticking points in the negotiations and articulate where the LDC Group stood on them. Next on my list to write were talking points for an evening event Marika had stuck in the chair's calendar.

Leaning forward during a lull in the debate, I caught Pa Ousman's attention, intent on getting some clarity. "For the event tonight, I've added in the LDC Group's positions on adaptation. There's a question about the national impacts of climate change. Is there anything you'd like me to put in about The Gambia?"

I hoped he would say no, because I didn't know what to write. I still knew remarkably little about The Gambia. I did know the country was on the Atlantic coast of Africa. Google told me that two-thirds of Banjul, their coastal capital, was less than a foot and a half above sea level. I sensed from Pa Ousman's and Bubu's dedication and the passion with which they spoke that climate change meant losing a great deal more than coastal real estate.

Pa Ousman must have seen the uncertainty in my face because he smiled. "No, that's fine. Just mark the place. I'll talk through that myself. Please make me a print. When is the event?"

I looked down at the Google calendar open on my laptop.

"Your calendar says 6:30 PM. I'll just double-check with Marika." I rose to leave, then hesitated. "Are you two getting lunch before the coordination meeting?"

Achala tended toward frenzied in her scheduling, and Pa Ousman didn't appear to value breaks with the same reverence that I did. Even knowing the answer, I waited until their heads shook in unison before calling Marika, who had set up shop in the LDC office. There was much to learn sitting in negotiating rooms with Pa Ousman and Achala, but it was nice to spend time with someone who regularly stopped for important things—like lunch and dinner.

Together in the buffet line, Marika and I caught up with Bubu, who shook his head when we paused to load our plates with salad greens.

"Why would you want to eat grass?" Bubu put on a comically confused face and tutted. I rolled my eyes.

Marika took a stab at persuasion, attempting a logical approach. "Greens are more sustainable, Bubu. Methane is thirty times more powerful than carbon dioxide, and do you know how much methane raising cattle produces?"

Bubu was too busy piling his plate to answer. In dining with the Gambian delegation, I'd come to understand that unless they had consumed rice and beef, eating had not taken place. Though fish could do, beef was the meat of choice. And rice was a must, the staple around which meals were built. Bubu acted like he couldn't go on if dining regularly consisted of other options.

After cashing out, we followed Bubu toward a half-empty table where Mbaye from Senegal sat eating something similar to the tower Bubu carried.

"Look at him," Bubu said waving at Mbaye's thin physique. "Look how young he looks. A meat diet is best."

"Exactly!" Mbaye laughed agreement, up to speed. This wasn't the first time vegetarianism had baffled our African colleagues.

Still, I had to try. "Just think of all those greenhouse gas emissions," I lamented, shaking my head. Global livestock did account for about 15 percent of the total.

"I am," Bubu said, shoveling a bite into his mouth. "That cow will never emit again!"

I couldn't help joining the laughter that echoed around the table. And really, The Gambia held essentially no responsibility for creating climate change. Bubu could make a compelling argument that equity demanded his right to all the beef he pleased. It was only fair that Marika and I, citizens of the fifth and first largest greenhouse gas emitters, should abstain. That wasn't the point Bubu was making, but I argued it through for him over the course of the meal.

Our conversation also mirrored exactly what was happening in plenary. We all agreed that we must stop the climate crisis. How we went about doing that, though, well—we all had different ideas about that.

After lunch, I printed a copy of Pa Ousman's talking points to hand over before his event. Curious, I followed him there rather than heading out to dinner with Marika.

I claimed a chair in the back half of a sparsely populated room and settled in to listen to the panel. All the action that goes on around the negotiations generally took place at events like these, or that was what I'd heard in Durban. I didn't get much time away to verify that. A PARTY badge wasn't required to attend, so everyone registered for the conference could go. At this evening's session, there were about twenty of us. The Bonn side event room was essentially a classroom with a table for the panel at the front and rows of metal chairs for the audience.

Pa Ousman took his place with the other panelists.

While Bubu freely offered up information about his life back home, my conversations with the LDC chair to date had focused on work—with little extraneous detail. The handful of Gambian delegates I knew, all tall men with bold tastes in fashion, added a bit of context. But I still didn't know much about The Gambia, apart from random facts gleaned from Google—for example, The Gambia was the smallest country in continental Africa. South Africa remained my only experience of the continent, which granted the same familiarity as visiting Anchorage, Alaska, would give to experiencing Panama City, Panama. That was the nearest equivalent distance the search engine threw up.

I wanted to know what home was like for them and what about climate change brought Pa Ousman here, fueled him through sleepless nights and thankless hours, made the prospect of leading a negotiating bloc worth the effort.

Once introduced, Pa Ousman spoke like a technician—one who can hold a lot of facts and figures in his head—raising his eyebrows over the rims of his glasses to articulate important points. I started touch-typing, keen to capture everything, then found it difficult to keep up. He rattled through quickly, without filler words or pauses, talking about sea level rise, just not in the way I expected.

"So the saline water moves toward the lowlands, affecting the productivity of the soil as it becomes more acidic. You have saline intrusion taking place."

I gave up and eased off the keys, focusing instead on the images that came together in my mind.

He spoke of a river, the Gambia River, around which his whole country lay. I imagined it, for the first time. The lazy brown banks of a delta flowing to the Atlantic, mangrove green. The sound of birds. People fishing. A two-dimensional river, as he described it, a dance between partly saline and partly fresh water, ebbing and flowing with the rains—disrupted by the changing rhythm. The rise of the sea, yes. But mostly the impacted seasons: extended drought followed by unpredicted downpour. They shifted the tides, erratically.

Gambians could find fresh drinking water in fewer and fewer places. They could plant crops in fewer and fewer places. I imagined farmers growing the main staple: rows of rice in square green fields, waiting for the influx of ankle-deep water now coming too fast or not at all, hippos ending up where they weren't supposed to. Mom used to tell me to chase rabbits that got through the deer fence out of her garden and to line the bottom with chicken wire. I couldn't imagine any fence that would keep out a hippo.

Lost in my imaginings, I typed only what stood out. *Disasters are occurring on a yearly basis—either flooding or drought. We are really being affected a lot. Agriculture is a problem.*

Where would they grow the rice? It was the African mainland's smallest country. There was nowhere to go, no reserve of arable land to move to. And with 1.6 million of The Gambia's 2 million people depending on subsistence agriculture to survive, a change of climate disrupted everything, and every mealtime would mark the loss. It was so clear, so powerful, this glimpse of Pa Ousman's world. I told Marika about it later that night, arriving in our hotel room after a summer twilight had settled over the market square.

"It was good to hear Pa Ousman talk about it," I said, putting my things down and rummaging through my half of the drawers for pajamas. "Afterward he told me it's the reason he works so hard; the lives of everyone he knows are already impacted by climate change. He sees it every day."

I paused, letting it hit me again. As was the case for all LDCs, the UN was The Gambia's best shot at getting leaders to take seriously the responsibility of stopping climate change. Hearing Pa Ousman tonight filled me with pride to work with them on such a worthy cause. "It was moving."

"I'm glad you got to hear him." Marika stretched her arms above her head. The book she was reading when I came in fell open across her lap. "Seems a harsh contrast to how little progress we've made at the negotiations."

"I know!" I scoffed. "I can't believe it's taken them nearly two weeks to adopt an agenda."

Marika agreed with a yawn. I also couldn't believe that, in a few days, when she and Achala would return to London and the Gambians to Banjul, I would still be here. Unbelievably, spending even more time at the UN was going to work out.

———————

I'd managed to get an internship at the UN climate change headquarters in Bonn lined up for after the negotiations. Another great, unpaid opportunity. Worth it, though. And I was right, finding cheap accommodation was much easier from Germany. I'd spent the past few evenings meeting potential host families about renting for a few months. One was willing to take a cash installment I could actually afford.

The room was in the home of a large extended family. Every morning, the grandmother woke her grandchildren—and inadvertently the entire house—with a resounding, "*Mein Liebchen!*" It echoed through the courtyard and set me upright in bed well before my alarm. This, combined with a location that was walking distance from the UNFCCC skyscraper, meant I was both eager and groggy when I arrived early for my first Monday of interning.

"Take the shuttle to the castle," the receptionist told me when I asked for directions. "It leaves in ten minutes from the stop just outside." She pointed me back through the glass. As if interning in Germany couldn't get any more surreal. Not only would I get to know the research work the UN secretariat did when countries weren't negotiating, I would do it in a castle on the Rhine River. I couldn't contain myself. I skipped through the summer sun toward the shuttle, determined to get as much out of this experience as possible.

Fresh from hearing Pa Ousman talk about climate impacts, I was glad to be interning with the Adaptation Programme. Understanding its work would give me an overview of the ongoing adaptation projects. I was tasked with updating the secretariat's database on those projects underway in the LDCs and assembling case studies of innovative technologies. I was also asked to look into how other multilateral bodies classified their adaptation work and, before the end of the summer, to present a range of options for the UNFCCC to consider when formulating its own categories. Oh, and could I also look into the nature of the cofinancing the LDCs attracted for their adaptation projects? Sure. No problem. Deep in documentation, I had mountains to do.

I was part of a cohort of interns from all over: Germany, China, Sweden, the United Kingdom, India, Portugal, Austria. Dubbed the Emerging Minds of Tomorrow, or EMOTs for short—because we were geeky enough to know that acronyms made everything cooler—we spent any free time playing what I was constantly reminded to call *football* rather than *soccer* on the castle's greens, socializing in beer gardens along the Rhine, and eating ice cream in Bonn's market square. In an attempt to further my ability to communicate, I did attend the German lessons the UN offered new staffers. I learned that *mein Liebchen* meant "my love" and how to say "I'll have sparkling apple juice rather than beer, please." Otherwise I picked up very little. Even so, I enjoyed the dubbed episodes of *Die Simpsons* that aired every night.

I also had a surprising number of interactions with the effervescent executive secretary of the UNFCCC, Christiana Figueres, as her offices were in the castle too. The only context in which I had seen her to date was at the top tables of the negotiation's opening and closing plenaries. Though Pa Ousman and Achala spoke highly of her and the secretariat she led, to me she was a public figure. Now, I saw her short brown hair and strong brows and eyeliner move through the corridors at the pace her workload demanded. I liked her self-coined "stubborn optimism," and hearing the boss laugh made for a nice work environment.

She called an interns' lunch during my second month. We sat cross-legged in a grassy circle, in a shaded spot by the river. Christiana wanted to know why we were passionate about climate change. When it was her turn, she told a story about golden frogs, her favorite animal as a child. Brilliant orange, the frogs would fill the dense green-black with bursts of color, bounding through Costa Rica's cloud forest into the dripping world above her home.

"Then I had little girls of my own. They were born in 1988 and 1989, so I imagine they're your age."

"Older." Someone laughed. Astounding colleagues with your birth-date was an unexpected perk of interning. I laughed too, although her daughters were essentially the same ages as my sister and me.

Christiana joined us in our laughter, then went on, "By the time I had my girls, the species no longer existed."

I went quiet. Growing up, my only reference for extinction was *Jurassic Park*. In real life, the word felt so sad, so wrong. How do you grieve the loss of an entire species, something so special, invaluable, never to return?

"It had a huge impact on me," she continued. "I realized I was turning over to my daughters a planet that had been diminished by our careless-ness. By our recklessness. That's why I'm passionate about what we're doing here." She ended and looked back at the offices of the secretariat she led. I remembered admiring her as a person and a leader—a combination I rarely found. The story also explained the frog illustrations on the wall of her office, which I had thought an odd choice at first.

My weekdays began with the walk to UNFCCC headquarters to catch the shuttle. The routine of passing through the castle's security registered as normal now that I spent so much time at the UN. My upstairs office was down a long hall. I entered every morning with a *"Guten Morgen"* for the woman I shared it with. As was the way of things, I felt just about settled as my time in Germany was winding down. While I waited for my computer to turn on and load my inbox, I thought about how strange leaving would be.

I wasn't in the habit of calling or writing my parents, so an email from Mom stood out. "Your dad" was the subject of my newest unread message. Mom wrote:

Your dad, at the present time, is having a bilateral lung transplant done at the University of Washington.

It stood there so alone, the sentence. Without context it took some time to sink in.

Having a double lung transplant was Dad's only hope of regaining a "normal" life. And lungs were only available should the worst happen to someone young and healthy. The five-year survival rate for the procedure was 50 percent. If your body rejected the new lungs, there was no going back. No plan B. I knew all of this, and so did he.

And unlike my mother and my sister, Dad had nothing to say to me.

I got up and walked out of my office, blinking back the onslaught of conflicting emotions. I knew Dad treated me differently. He always had, but it grew increasingly pronounced after he'd returned from Virginia all those years ago. Christmases during my adolescence were spent watching him give presents to my sister and not to me. Chanteal's middle school performances and sporting events were attended, while mine were not. And he didn't hit the other members of my family. Nor would he neglect them to the extent that he would me. Even so, I never thought that an email from Mom could well be the last I heard of my father.

I called Chanteal and got a brief recounting of their morning's events: the doctor's call coming through, Mom and Dad leaving for the hospital.

I called Mom.

"He's recovering from surgery." She sounded harried, like she was focused on too many things to organize her thoughts. "The doctors say it was a success, but we'll see how things go. I need to go check in to the health-care facility soon."

Mom would need to stay with Dad in a Seattle facility for the next several months. UW hospital was the only place in the region where transplants were done. Recovering patients had to stay in the area until they were given the all clear.

She had gone quiet.

"Chanteal says she'll come up to Seattle this weekend," I said.

"Yes, she will," Mom responded.

"And I'll call Michelle and Andrea. They're probably at the hospital right now too," I added.

"I don't want to bother them."

"They won't mind," I said with certainty. "They wouldn't want you to be alone."

I pictured Mom sitting in a hospital waiting room, the chairs with ugly patterns and light wood trim. Muted music, stale coffee, and old magazines. Alone. Mom was sitting there alone. My throat tightened and my eyes stung with the threat of tears.

She remained quiet.

"I should probably go," she said at length.

I couldn't think of anything else to say to keep her on the line.

"OK," I answered. "I've sent you my German number. I'll call you tomorrow." The line went dead before I finished speaking. I didn't know how much she heard before she hung up.

I was in the break room on the floor above my office. The windowless space had a broken printer on one side and a table tennis table on the other. I didn't turn all the lights on when I came in, so the far corner where I sat on top of a dusty filing cabinet was dimly lit. I leaned my head back against the wall and closed my eyes. I didn't know what to feel.

My father had undergone life-altering surgery without a word. In death or life, it seemed he had nothing to say to me. I wondered who Mom would be when I saw her next, after spending every waking moment with him for the next several months. Perhaps she would retreat so far within herself that I wouldn't know her either. Tears rolled down my cheeks. If I lost them both, who would I be?

I wished someone had told me.

I wished things were different.

Yes, I could run away—run the world away. For so long my only thought had been to leave, to get as far away as physically possible and stay there. Travel, move to Rhode Island. Build a life for myself, one that was safe. One that was free. And I was good at it. Bonn was over five thousand miles from Kelso, where the rest of my family still lived and I remained adjacent. Knowing that had always hurt less than staying with them, or at least, I thought it had.

Mom was sitting in that hospital alone. And Dad had nothing to say to me.

God, it hurt.

My chin bent up in pain, past the point of tears. I didn't want this. I wanted more, better than silence in the face of life-altering surgery. Better than a final chapter spent not hearing. Nothing like death to put things into perspective. Those two people at UW Medical Center were the only set of parents I would ever have.

I stared at the cell phone waiting almost expectantly, wandering around in my mind. They wouldn't call. They never called. The typical intervals between our conversations were measured in months, not days. And it wasn't just them. After I moved out, Mom resorted to emails asking if I was still alive, I let it go so long.

Maybe I could call.

Even the thought had me drawing back, putting the phone down. I left the break room resolved to give it twenty-four hours and see. I couldn't. Couldn't sleep either. Attempting to not think about it was a hopeless waste of time. The puffy eyes that greeted me in the mirror the next morning looked like they'd taken a beating. All the moments of wavering resolve, the untested possibilities. All right, fine, they were there and I was here, my time in Bonn unfinished. What's the worst that could happen? They'd hang up?

I would call.

So I talked to Mom, asked how things were going, how she was, what the doctors had to say. The day after that, I called again. What Michelle and Andrea said. How he was doing. And it was nice. To know. To hear.

I could do this. I could call.

Dad's hospital stay sped by at a breathtaking pace. Less than a week all told. Then he and Mom moved on to the Seattle health-care facility, as was policy, remaining close to the hospital without occupying a bed in it for the next several months. Their treatment center was in South Lake Union. This meant that whenever Mom went out for groceries, she passed tech execs and tourists wearing RIDE THE SLUT T-shirts.

"I can't believe they walk around like that," Mom told me. I still marveled at talking to her unannounced, though Mom had owned a cell phone for years. It was the ability to reach her with it that was new.

Until recently, she had adopted the annoying parental habit of turning the cell phone on only when she wished to call someone.

"It is pretty funny," I said. "But then, I lived in Seattle, so I know that it stands for South Lake Union Trolley." The streetcar's reputation well outpaced its few miles of track connecting the wealthy neighborhoods south of Lake Union to downtown. "I guess it would be more shocking if you didn't know that."

"I'm back now," Mom reported, panting. She had walked uphill several blocks and then upstairs to their room. I understood from descriptions and the pictures Chanteal sent that their room was a cross between that of a hotel and hospital, more comfortable sterile furniture and more space for slightly more permanent stays. As Mom settled herself back inside, she passed the phone to Dad.

"Hello," he said.

"Hi."

It was still weird. Though I had spoken to my father on the phone before, we did not chat. Our conversations had always held a specific purpose: some tangible outcome or question that needed answering. Without that structure, I floundered, completely unsure.

"How's it going?" I asked into the silence, reminding myself to try, that I wanted something different. "How are you feeling?"

"Only one chest tube left in. It's amazing how quickly the swelling's gone down."

He sounded so different. The role lungs play in verbal communication astonished me yet again. The mechanics of artificial breath were entirely gone. He didn't sound like Darth Vader at all. I had taken for granted the silent influx of air, the force supported speech registered. The difference to the actual sound of his voice was remarkable.

"They have me walking the corridors every few hours," he went on, unencumbered. "Just short distances, but I can see the improvement."

And there was the biggest transformation. Dad was being positive. For the first time in a decade, his health was steadily *improving*. Optimism changed his demeanor. I bet if I saw him now, I wouldn't recognize him. Maybe there was a Skywalker in there.

"Wow," I marveled. "How's the indigestion?"

Yesterday's conversation brought up that his stomach spiked with acid whenever he swallowed. Adding food made for an intolerable amount of reflux. Not eating would suck.

"No change, so we're experimenting with different foods."

I could hear Mom unwrapping a variety for lunch in the background.

"The doctors think it will go away once things settle a bit," Dad said. "Focusing on the lungs now."

"When are the doctors coming today?" I asked.

"Supposed to be here in the afternoon. I'm meant to eat something first and give them an update."

"Well," I said. "Good luck with lunch. I'll talk to you tomorrow." Because that was what I did now, I called my family every day. It was bizarre, but it was the more I wanted, this unknown level of familial intimacy.

I heard months of test results and progress reports, which informed moves between the hospital and the health-care facility and eventually from the health-care facility back home to Kelso. All told, the transplant was hailed a success. Dad would live an oxygen-saturated life, free from the restrictions that had kept him dependent on tanks and access. Free to resume the lifestyle he thought had been taken from him. Free to live happily. And he seemed so *happy*. Our conversations passed without incident frequently enough that I stopped having to remind myself to try.

Dad laughed without agenda, tolerated differences of opinion. He asked questions about my life whose answers did not directly affect him. By late August, I decided to visit. I had to see this for myself. Besides, since his surgery, my mind had struggled to remain in Bonn. My internship culminated with a shaky presentation, complete with half-finished responses to the questions put to me. I seemed unable to concentrate with any depth on anything other than the change I might find back home. Still, I underestimated it.

Dad looked better than I imagined, better than I could remember. Even his color was better. After ten years of seeing him cough, wheeze, and pack around oxygen, just watching him deeply inhale was a spectacle. I followed him to the track and the grass fields surrounding Kelso

High School. Both awe and trepidation overwhelmed me as I watched my father, a man who could not run, sprint with steadily lengthening strides.

He was a different person.

And we were different with each other. We had never spoken so much. Not since childhood, well before I moved out of his house. Sure, I was always the one to call home, but it felt like the effort was paying off. We talked nearly every day now.

I stayed at my parents' house, slept there rather than at Noelle and Kevin's. I dared to think that my family had turned a corner, wherein I could imagine a new future for myself. What if the threat that had kept me at a careful distance didn't exist? What if I didn't have to run away? Perhaps this breathing, smiling man would love me. Perhaps we could have a different relationship.

I couldn't stay long. My Bonn summer had left tons of thesis work unfinished and waiting for me back at Brown. When my time in Kelso dwindled, I ran to the end of our road to stare intently at Mount Saint Helens's steaming summit, set against the clarity of the autumn sky. The air was so nice to run through, crisp and clean. Back home, I stopped in the kitchen for water in the stillness of my parents' empty house—a pause that coincided with Dad's arrival home. At any other point, this would have triggered me to make myself scarce. But things were different, so I waited. Stayed to talk.

"The thing we need to focus on now is how fat your mom has gotten."

I leaned against the wooden back of the chair, arms folded across my chest. My legs splayed out encased in slowly drying, muddy track pants. Beyond the line of trees outside, I could see the sunset begin to color the sky through the wall of windows. Dad looked at me from across the kitchen table.

I willed myself to have misheard. "What?"

Dad had come from the gym, which was apparently something he did. The physical therapy he started at the Seattle health-care facility, combined with his newfound mobility, was growing into a near-religious emphasis on daily exercise. I still registered shock that this was even possible.

"Your mom, she's been overweight for years," he confirmed.

I stared at him while he went on, watching my illusions vaporize. Dad was the survivor of a double lung transplant, a miracle of modern medicine.

And still a giant ass.

That surgery could have so easily killed him. He's the only father I had.

I thought these things over and over while he talked, trying to decide how much I trusted our new relationship, how much I was willing to give.

"If you're so concerned about it," I said carefully, "why don't you work out together?"

He laughed. "Your mother won't work out with me."

"Have you asked her? It seems kind of perfect. You're just getting back into it. She hasn't been in a while . . ." I trailed off.

"I've tried that route with your mother," he said. "She needs to take responsibility for her own health problems."

Of all the hypocritical bullshit. I had watched my mother take care of my father for well over a decade. And so had he. The diplomatic line I was trying to tread evaporated in my exasperation. I couldn't do this.

"Mom spent months in Seattle looking after your health. Why don't you devote the same amount of time to hers? It's not all about you, you know."

That did it.

I watched the contempt spread across his face, saw all the weeks and months of relationship building go up in flames.

The shouting began. "You have no fucking idea what you're talking about!"

He was up from the table. I saw his chest heave with oxygen, expanding rapidly in and out—still marveling. Only part of my brain registered the fear and knowledge that coughing fits would no longer hamstring him. My body tensed, frozen. I waited for him to lunge.

He turned away, stalked off down the hall, leaving me blinking in disbelief.

To his credit, I had no idea how my parents' marriage worked.

———————

My last full day in Kelso passed in silence—at least from him. I took this with resignation. Perhaps talking every day was beyond us. Our final family dinner was an awkward affair of Mom, Chanteal, and me laughing as Dad's bitterness saturated the air. The next morning, I rolled my suitcase into the open garage in preparation for Noelle's arrival. Even with family relations at this high point, Noelle was still the one taking me to the airport.

Ahead of schedule, I perched on the edge of my suitcase and prepared to wait.

The noise of a door opening made me look back at the house, where I was surprised to see Dad standing in the frame. As a rule, my father did not say goodbye to me, a custom I assumed I had rekindled with our last conversation.

"Brianna," he addressed me.

I swallowed, secretly hoping another Brianna would appear. "Yes?"

"I have something to say to you before you go." He had clearly thought about this. "I was astonished by what you said the other day," he began.

I didn't voice that I felt exactly the same.

"It's clear that you base your opinions on beliefs, rather than facts. This is a new side I'm seeing of 'Brianna,'" here he stopped to raise air quotes around my name. A large part of me struggled not to laugh. "One that I don't approve of."

Shit.

"You should never say to a parent, 'it's all about them,'" he continued. "I have given you everything." He paused to let this sink in. I held back the range of colorful responses that came to mind, numb to everything but indignation. "Fortunately, though, you're not a child anymore. And like any stranger I disapprove of, I no longer have to deal with you."

The list of retorts I was busy choreographing all balked in response. I was suddenly half my age. The parallels between the two conversations recalled a memory so vivid I unwillingly relived it. I was thirteen again.

My middle school friends were donating their hair to Locks of Love, cutting off long ponytails and going for short dos. I wanted to join them. A friend's mom would pick me up. Neither Mom nor Dad had said no, so I thought their consent was implicit. The next day, I waited, with a freshly shorn afro, for Dad to pick me up outside Coweeman Middle School. I had an appointment I wasn't looking forward to. I didn't enjoy the discomfort of orthodontia. When Dad arrived, I opened the passenger door as usual.

"Let's have you sit in the back today."

Confused, I examined the front seat. There wasn't anything in it. He just didn't want me in it. This had to be about more than just my hair. I thought it looked good. My reaction was to spend the fifteen-minute drive deftly ripping two eyeholes into a brown paper bag I found in back. Once we arrived, though, I didn't have the guts to wear it. When we returned to the school's parking lot after my appointment, Dad made his feelings clear.

"You are no longer my daughter." Like my lineage was something he had changed his mind about. Like that was a decision he could make. I stood outside his car window, blinking, unsure of what that meant.

As I did now. Twelve years later the words "I no longer have to deal with you" echoed in the space between my ears—crowding out all other thought.

I would never be enough.

Light flashed in my periphery as Noelle's headlights brimmed over the curve of the driveway. Seeing me and my suitcase waiting, she turned and reversed in a practiced movement. The pop of her trunk broke my stunned silence.

Having said what he came to say, Dad headed inside.

The "Goodbye" I offered in parting hung unanswered in the driveway.

6

COMPROMISED

Providence, Rhode Island
September 2012

BACK IN PROVIDENCE, I moved into a shared apartment across the street from Becca's place. From my front room window, I could see Becca's back window in the property fewer than a hundred yards away. It took exactly sixty seconds to get there. We worked this out in the hubbub of starting our second and last academic year at Brown University.

What remained of my graduate education was the research of an independent thesis. I was naturally a night person, so with no class schedule to wake me, as autumn stretched on I began a gradual shift into a bat-like routine, which Becca happened to share. The small cohort of the environmental studies graduate program lent to building close relationships, and Becca and I now spent an inordinate amount of time together. As the months passed, we developed our own routine.

We texted every "morning," meaning around noon, to make a plan. This usually involved huddling over laptops to write until the hunger kicked in and one of us asked, "Dinner and *Dawson's Creek*?" Rewatching that particular show became our slightly unfortunate and very entertaining habit while stir-frying dinner, after which we tried to convince each other to keep being productive and either Becca or I would head back to our respective apartments to work until the early hours of the morning.

This meant that on September 23, at around 11:00 PM, I turned off the computer and headed for my darkened bedroom, where I curled into a ball on a cushy green armchair. Today the fact that, at age twenty-five, I still didn't have a functioning relationship with my father had entirely dominated my thoughts.

"Abba?" I said this aloud, as was my habit.

God, my Father. Not to be confused with the pop band. While I was in undergrad, an eloquent preacher enraptured me by describing the Holy Spirit as a friend that constantly surrounds us, whose unfailing love allows us to cry out, "Abba!"—an endearment for father, so "Dad" or "Daddy"—and expect a response. A girl with a father such as mine latched on to that hope with tenacity and didn't let go.

"My love," this conscious tone ran through my mind in greeting.

It was distinct from what I considered my own thoughts and imaginings. It was audible only once, at the height of childhood trauma, and held in reverent skepticism ever since. The worry that losing my mind was a real possibility meant very little then. A neglected teen, I was too desperate for affection to let it go. If hearing voices was the price of having someone navigate the complexities of my world with me, I would pay it, gladly. Afterward His voice came like a radio channel my mind was attuned to hear. Sometimes the signal was weaker than others. Sometimes there was radio silence. And sometimes I couldn't be sure of what I heard.

But not tonight.

I can't shake it, Abba, the sadness, I thought these words in response, like a conversation. *Every time it goes quiet, I just sit,* I went on. *I don't like thinking about what's wrong. When I remember, it's worse.*

I exhaled.

Please help me, I prayed on. *I haven't told my friends. Would that help?* I didn't want to talk about it though. Who starts a conversation with, "So my dad's not speaking to me—again"? The whole point of leaving was to escape this dynamic. I didn't want to feel this way. But I couldn't avoid this part, the sad, mopey blankness, the inability to do or focus on anything else.

I just don't know what to do. I exhaled—exasperated, wrung out. *I'm hurt and I don't know how to heal.* Admitting. Acknowledging. A moron for trying. I should have known that I would never be enough.

How do you replace a father's love?

My stored-up torrent done, it was finally quiet in my mind. A moment passed before I heard, *"With another's."*

Today is my father's birthday, I thought, coming to the point. The date marked over a month since we'd spoken and made our broken dynamic unignorable. *Abba, what do I do?*

"I believe it's typical to call."

I rolled my eyes in response. *Thanks,* I thought sarcastically. Abba was rather funny. I supposed being eternal granted enough perspective to find humor in just about everything.

What do I say? I clarified. *And if You say, "Happy birthday," this conversation is over.*

"Brianna, what's the matter?"

It hurts too much, Abba, I thought angrily. *It hurts to love someone who doesn't love me, who thinks our relationship is nothing, worthless, like that of a stranger.*

I was mad now, angry that, despite a lifetime of evidence, I still wanted something I would never have.

He. Doesn't. Love. Me.

And then I started to cry.

We sat for a while, until the sobbing eased up.

"He thinks he does," I heard. *"And you know love is not a reciprocal act. It's an independent choice that I am encouraging you to make. You've asked what I do all day. Showing love to those who don't love you will teach you more about my 'days' than anything else."*

How sad, Abba.

"I don't find it so—not in the way that you think. I am Love. It is not difficult for me to be."

Doesn't it hurt when we don't choose to love You? I asked.

"Yes, but you cannot yet understand how. It is not the same pain you feel at your father's dismissal."

I want to give up.

"Then do."

Again, with the perspective—so unhelpful. I wasn't finished. *Why am I so selfish? So ungrateful?*

"I think you give yourself too hard a time."

I wish I were You, I thought.

"Don't wish that. You'll go straight to hell." I visualized Lucifer falling from grace in a flash of lightning, and the absurdity of the comparison made me laugh out loud, undoubtedly by design.

Ha ha, Abba, I articulated in my head. *Always so witty.*

"Why do you wish you were Me?" He asked me to clarify.

Because then this wouldn't be so hard. Every choice hurts. If I don't try, I lose my family. If I do, I lose myself, I thought in defeat at the impossible situation. *And the biggest problem of all is that I love him. He's the only father I'll ever have.*

My lips mashed together in a tight line, threatening a fresh wave of tears. This remained the hardest thing to admit, the hardest thing to understand. I wanted him to love me, to approve of me. I always had. And I couldn't earn it. I certainly couldn't keep it. I was ashamed to admit how much that destroyed me, even to myself. *Where do we go from here?*

"I suppose you call."

I suppose I do. I sighed, relieved at the decision finally made. *Abba, I love You.*

"I know. And you know what? I love you too."

Only more.

His *"Of course"* made me laugh again.

I got up and crossed the room to locate my cell phone where it was charging on the nightstand. For the past several weeks, if I called the house and Mom wasn't there, Dad didn't answer. I pressed the appropriate buttons. My parents' landline, unchanged my entire life, remained the only number I still had memorized. It was nearly midnight. Almost 9:00 PM back home. I listened to the phone ring, counting. The third was followed by a slightly longer interval. It heralded my father's voice on the answering machine: "You have reached the Craft residence. After the tone, please leave a message and we will return your call." BEEP.

To the expectant silence, I said, "Hi. This is Brianna. I guess you're out. I was calling to say happy birthday. Give me a call back when you get this message." In the still of a Providence night, I hung up. I knew he would not call back.

Over the next several months, the northeastern leaves changed with a blaze of color. I watched them fall, ruminating on the false sense of security the ease of the atmospheric exhale instilled. I typed messages to a team of climate change negotiators half a world away. I planned and eventually packed, all while waiting for a resolution.

———————

"Where are you going again?" Noelle asked over speakerphone. Through my slightly ajar bedroom door, I heard the whistle of the kettle my room-mate was boiling; at nearly the same moment, the baseboard radiators hummed to life. I shivered out a smile. The heating coming on was the highlight of winter afternoons.

"Qatar—Cutter? I'm not sure how to pronounce it." I was folding. Clothes lay scattered across my high-ceilinged, robin's egg blue room. "It's a little peninsula off of Saudi Arabia," I went on. "I'm excited. I've never been to the Middle East before."

Then, looking at the mess around me, I frowned. Packing long-sleeved, long-skirted business wear wasn't going well. Potential outfits covered the surface of my twin bed as well as that of the cushy green armchair tucked into the room's corner. They ran over my open suit-case and down onto the dark wood of the floor, in a frustrating array of half-completed looks. Turns out, I didn't own much that was both suited for hot weather and covered ankles to wrists, which guidebooks informed me was the appropriate attire for women.

"I hope you like Qatar," Noelle wished through the cell phone propped up on my nightstand. "Take pictures. I'm sure it will be interesting."

"I hope it's not *too* interesting." With two negotiations under my belt, I certainly felt more prepared for my second COP than I had my first. Even so, I was hoping for a descriptor closer to "uneventful" than "interesting." I could do with some uneventfulness in my life. A trip to the Arabian Peninsula probably wasn't the best place to start.

From thirty thousand feet, I gawked at the moonscape. Miles and miles of what looked like a vast, uninhabitable expanse of sand stretched between our plane and the Persian Gulf. This was suddenly broken by a metropolis springing from the dunes. My jaw dropped as we touched down in Dubai on layover. The sight of the world's tallest skyscraper, Burj Khalifa, winding a third of a mile up left me feeling wonderfully small and insignificant. I marveled while happily contrasting the evening temperatures of eighty degrees Fahrenheit with the freezing ones I had left behind in Providence's winter.

Yet again I arrived at my final destination with instructions from Marika. A steady flow of emails had kept us connected since we'd said goodbye in Bonn. Most of them related to doing odd tasks for Achala, but we interspersed these with jokes and funny news stories about British-American relations. When Achala and Marika started planning for COP, I made the same housing plea as I had six months prior. Though I would love the opportunity to work for the LDC Group, I remained a broke grad student. Happily, Marika offered to share her hotel room in response.

After sipping tea through check-in at the multistoried hotel in Doha's south side, I hugged her in hello and thanks in a room three times the size of the one we shared in Bonn.

"Look at this bed," I said giddily, gesturing toward one of the queen-sized mattresses. "What will I do with all this space?"

Marika moved across the room with arms outstretched, a smile on her face. "I'm sorry. I can't hear you from the sitting area," she called from the lounge chairs surrounding a small table on the other side of the room. "You'll have to speak up." We spent most of the morning laughing over the comparison.

Our hotel was across the street from a mosque and a small park. Several times during the day, the call to prayer rang out from minarets across the city—the steady, solemn stream of recorded Arabic calling in the faithful. I watched from our balcony as the mosque overflowed with men who moved to form neat lines in the park, all kneeling and bowing in unison as they prayed toward Mecca. I observed with curiosity, thankful that our fifth-floor window featured such a nice view of a religion I had never seen in person.

Outside the thirteenth-story window of the breakfast room, the city of Doha stretched around the crescent of a bay. Wooden fishing boats bobbed gently in the turquoise water, tranquilly contrasting the formidable line of skyscrapers crowding the northern shore.

"It has to be empty, right?" I asked Marika again, waving toward the cityscape.

The skyline mirrored that of New York's, only the population was fewer than a million people—an eighth that of the Big Apple. I had googled this yesterday, unable to reconcile the breadth of the built environment with the scattering of people I saw on the street.

Distracted by the buzz of a text message, Marika just shrugged in response. "Achala says she'll meet us at the prep meeting."

"Oh, OK. I'm ready if you are." I smiled.

The taxi wound its way around the waterfront toward the northeastern end of downtown Qatar's vast and vacant skyline to the building designated for the LDC Group's preparatory meeting. The cars on the road were large, American-style models that glittered in the sun. Only a few people braved the soaring temperatures of the sidewalks lined with potted palms. I scoured the levels of the skyscrapers as we passed, searching for life. Given the relatively few people, I wondered if you could easily rent entire floors in one of the several dozen buildings on display. The emptiness of downtown left me oddly unsettled and conjured up images of a strangely vacant approximation of an adult Disneyland.

When we eventually disembarked, I took comfort in the familiar faces that waited for us beyond the security lines. I was keen to get to know Pa Ousman and Bubu better this session and had done some preparations of my own.

"Biryani!" Bubu called as I came through the doors of the meeting room.

"Bubu!" I yelled back, only then registering what he had said. I wasn't his favorite Indian rice dish. "Did you call me biryani? Not everything is food, Bubu!" I smiled, falling back into the easy rhythm we had developed over the past year.

He expelled peals of laughter accompanied by his trademark belly pat. Apparently, I had some more work to do to imprint my name in Bubu's mind. Once he collected himself, he cried, "Ay! These American names: Bri-hann-na. They are so complicated!"

We laughed together.

"How are you?" he finally greeted. He paused for only the briefest of seconds before following it up with, "Do you have something for me?"

I laughed again, knowing what he wanted. I learned in Bonn that Bubu shared my love of chocolate, so much so that telling him that I usually carried a bar in my purse was a big mistake. Giving Bubu "some" of your chocolate meant giving Bubu all of your chocolate. During a slow LDC coordination meeting, I watched him eat through an entire box of truffles in a single sitting. Where Pa Ousman always asked for Coke when I offered drinks, Bubu now only asked for chocolate bars.

I shook hands with Pa Ousman and pulled something out of my purse for him as well. "I thought of you when I saw these in the airport. They're a reminder to take some breaks." I handed a few *People* magazines to him.

A smile spread across his face as we locked eyes. I had always known Pa Ousman was an avid reader. I was usually the one tasked with printing out emissions counting methodologies or articles about international relations for him. In Bonn, though, I had also picked up that Pa Ousman's interest areas extended to American celebrity gossip—discovered when, late one night, I saw a copy of *Vanity Fair* spread across the LDC office table. He laughed briefly before collecting himself.

"Thank you, Brianna. Knowing these things is important." He nodded seriously, already thumbing through the glossy pages.

Achala—who hugged me now in greeting—joined in Marika's laughter. It was nice that the range of our conversation could now cover more than upcoming meetings and clarifying the group's positions. Unlike with Bubu's open temperament, getting to know Pa Ousman was more challenging. I was glad my effort had paid off.

Pa Ousman called the room to order once enough LDC negotiators had filed in to fill its tables with country flags and its chairs with suits. As the two days of discussion got underway, I tried to discern what would need to happen for the negotiations to make more progress here in Doha than they had in Bonn. Three things bubbled to the surface of my notes: extending the life of the Kyoto Protocol, setting up the Green Climate Fund, and working on the new treaty. The LDCs remembered the deal they'd made in Durban: the first two in exchange for undertaking the third.

The Kyoto Protocol—the existing agreement wherein developed countries pledged to reduce their greenhouse gas emissions—would end in a month. In South Africa developed countries agreed to extend their Kyoto commitments while we worked to negotiate the new treaty, under which everyone would make cuts. LDC delegates now discussed how long to extend the Kyoto Protocol, agreeing that anything longer than five years had the potential to lock in inadequate rates of emissions reductions. A "second commitment period" of five years would end in 2018, leaving time to review progress and recommit prior to the new agreement coming online in 2020.

"Let's see the pledges developed countries put forward," a delegate said, "before we implement our own. Let's see if they are serious about reducing emissions before we commit to reducing ours."

In Durban negotiators had also agreed to bring to life the Green Climate Fund. Since then, some progress had been made. The Republic of Korea would ask the COP next week to endorse its city of Songdo as the home of the fund. But the money promised to fill the Green Climate Fund remained unforthcoming. Here again, flags went up around the room.

"Where is the money developed countries promised? How can we negotiate new deals if they are not delivering on the $100 billion?"

Pa Ousman nodded in agreement. "What we need is a road map, a clear road map that lays out how much they will put forward every year." The LDCs would expect agreement on a plan that mapped the annual increase of public finance contributions needed between now and 2020—by which point developed countries had promised the total would reach $100 billion per year.

"Should there be strong commitments on finance and an extension of the Kyoto Protocol, we will be in a position to make progress on the new treaty," Pa Ousman said, moving the discussion forward. "According to the agenda agreed in Bonn, we will divide the work into two parts." Bonn's agenda fight had culminated in the development of two work streams—developing a post-2020 regime and working on pre-2020 mitigation. The stream dedicated to the post-2020 regime would shape the new agreement. The other would focus on getting more emissions cut before the new treaty came into effect, namely from the developed countries that had shed their Kyoto Protocol commitments before Durban—the United States and Canada. Some developing countries would also need to start making reductions pre-2020 if we were to have a chance at stabilizing global temperature rise. A massive amount of work lay ahead.

At the end of the preparatory meeting's second day, Achala, Marika, and I followed Pa Ousman out to the waterfront. Our exploratory ramble ended with an impromptu boat ride around the bay. I breathed in the heat and watched the setting sun burn a red hue against the buildings, their lights slowly illuminating the darkening sky.

One of the things we needed to locate was the conference center meant to hold the negotiations. It was set to open on Saturday. Marika and I traveled in together to get our bearings and find the LDC office. A shuttle waited right outside our hotel, which we boarded optimistically.

"Getting there will be easy, then," Marika said as we sat down. When an hour later we were still driving away from the city, I felt her optimism subside.

"Where are we going?" I wondered aloud.

The terrain outside the bus had slowly transitioned from the skyscrapers along the waterfront to a sparse, sand-blown landscape. When the driver finally pulled off the highway, a gleaming conference center that looked like it had been constructed for the occasion came into view. I thought it oddly located, in all the emptiness. The building was so vast it recalled the long tentacles of American airports, a comparison exacerbated by the presence of moving sidewalks. I disliked the extreme difference in temperature between the 100-plus degrees Fahrenheit exterior and the cardigan-inducing chill of the air-conditioned interior. But my least favorite part of the complex had to be the sculpture that loomed in the central foyer: a sixteen-foot arachnid, complete with an egg sac hanging from its abdomen. People were walking under it. Yuck.

The Qatari delegation of white-clad men took the stage to address the world at the opening ceremony. Christiana Figueres, veiled in a beige, bedazzled scarf, stood among them. The COP president, Minister Al-Attiyah, was a man with a white mustache who stammered when he spoke. He shook hands with South Africa's Minister Nkoana-Mashabane, relieving her of the role of steering the negotiations for the past year.

"Colleagues," he boomed too close to the microphone. "We have before us the challenge of seven negotiating bodies convening in Doha. We must agree to a second commitment period under the Kyoto Protocol and achieve progress on work undertaken in Durban."

From behind Pa Ousman, I clapped with the crowd when he finished, settling in to hear the different groups' expectations for Doha. My fingers were poised over my laptop, ready to capture how their statements squared with the questions and proposals the LDC Group had discussed last week. I was also intent on testing my knowledge of who's who in the negotiations. Knowing which countries worked together was key. I was glad that the Group of 77 and China went first. The largest of the blocs, they were given priority. Their meetings and positions had the power to shift the tenor and tempo of the negotiations. Literally, scheduled negotiating sessions would not begin until a G77 representative arrived. They spoke first on protocol.

The developed country groups were next. The United States participated in the Umbrella Group, which was a loose coalition of countries that most often consisted of the United States, Australia, Canada, Japan, and New Zealand. Then the European Union, whose twenty-eight members spoke as a group.

I liked the lead representative of the Environmental Integrity Group because he always listed its countries in his opening sentence, "Switzerland, Mexico, Liechtenstein, Monaco, and the Republic of Korea," and closed with an enthusiastic, "Thank you very much."

The African Group and AOSIS I remembered from Bonn. I reminded myself that Small Island Developing States were defined by UN criteria, like the LDCs, and AOSIS was the negotiating bloc that represented most island nations.

Then came the South American constituencies, which were the groups to watch for high drama. A negotiator of theirs was prone to standing on tables to get the attention of the presidency. The four major developing countries—Brazil, South Africa, India, and China—came together as BASIC. In sum, I listed thirteen blocs take the floor to speak. I was happy that I had to google only a few. Even as a COP sophomore, naming the members of each bloc was still difficult because so many constituencies overlapped.

What I still found difficult to understand was the succession of opening plenaries. After Minister Al-Attiyah's gavel went down, the placard of one chair after another replaced his, the seven negotiating bodies he referred to, each named with an acronym. I still prayed no one would ask me to spell these out. While interning over the summer, a common prank was to tell new joiners that they would be quizzed on the meaning of all UNFCCC acronyms at the start of their second week. Most found the prospect terrifying and would laugh in relief when someone gave up the joke.

I quizzed myself now. The body that Minister Al-Attiyah chaired was the Conference of the Parties (COP). Then there was: the Conference of the Parties serving as the meeting of the Parties to the Kyoto Protocol (CMP); the Subsidiary Body for Implementation (SBI); the Subsidiary Body for Scientific and Technological Advice (SBSTA); the Ad Hoc Working Group on Long-term Cooperative Action (AWG-LCA); and the Ad Hoc Working Group on Further Commitments for Annex I Parties under the Kyoto Protocol (AWG-KP). The negotiations in Bonn had dubbed the newest body the Ad Hoc Working Group on the Durban Platform for Enhanced Action, now known as the ADP. It would take forward the discussions pertaining to the new agreement.

Each of the seven bodies had its own chair, who set its agenda. Over the next two weeks, each item of each agenda was negotiated in a room somewhere in the conference center. And those agendas spanned the range of thematic issues. There was no one place in which technology was discussed, for example. I had to comb through all seven agendas and highlight each of the items I thought relevant to follow that one theme. It just didn't seem understandable, let alone logical. I wondered how negotiators knew where to start. How were seven negotiating bodies meeting simultaneously? And where were the decisions made?

While in Durban, I was too overwhelmed to try and comprehend a structure; I now made a concerted effort. I studied the agendas and meeting schedule, trying to see how the secretariat arranged the chaos into order. Week one of the formal proceedings seemed reserved for the nitty-gritty aspects of the negotiations. The least controversial agenda items were called first. Monday afternoon convened SBI agenda item 17, "Administrative, financial and institutional matters." I didn't attend, but when I asked Achala what would be discussed she waved the question away.

"A few countries will go and approve the representatives that have been nominated over the course of the year. It's just procedural."

With the help of the secretariat, the chairs of the seven bodies tried to clear as many agenda items as possible within the first week. As procedural and technical pieces finished up—some of them only needing to meet once for nations to sign off on the way forward—more politically challenging agenda items were introduced, like who was going to pay for what.

On the first Thursday of the COP, I trailed Achala and Pa Ousman into a session on long-term finance, eyes peeled for Evans, the LDC Group's finance coordinator. Evans was a middle-aged man from Malawi, a small country in southeastern Africa separated from the Indian Ocean by Tanzania, Mozambique, and Lake Malawi's long sliver. Though he always took a long time to speak, I found following his logic difficult. He also often laughed when he spoke, which confused me further, as the issue he negotiated was so hotly contested. Money usually was.

I spotted him sitting at the far end of tables arranged in a square. He saw Pa Ousman approaching and saved the empty chair next to him. Pa Ousman took it with a handshake while Achala and I claimed free seats in the rows behind them.

The facilitator called the session to order, and flags went up around the steadily filling room. I concentrated. Listening to good negotiators was eye-opening. No one said what they meant, and everyone was so overtly polite that it was difficult to follow the discussion. After taking notes for a couple of hours, I started typing out what I thought was actually being said rather than capturing the dialogue verbatim.

The facilitator read out that Norway was next to speak.

A man in a tie started with, "I would like to thank my distinguished colleague from Bolivia for that last enlightening intervention."

I typed, *Norway wishes Bolivia would shut the hell up and thinks that what they last said was dumb.*

Norway went on, "I also agree with the earlier intervention made by the Philippines: a nation's willingness to contribute to global solutions defies categorization and must be encouraged . . ."

I thought for a moment.

Norway thinks developing countries need to pay for climate change too. The Philippines didn't say that. But saying a developing country said this makes Norway's argument stronger. Norway hopes you weren't paying attention to the Philippines so you won't be able to object.

The best negotiators had a remarkable ability. Not only could they persuade people to agree to their proposals but good negotiators could also convince others that the proposal was their idea to begin with.

On some topics, however, even the best negotiators couldn't broker compromise. In Doha the late nights began before I anticipated. I thought Durban would have been an exceptional COP given the stakes of the topic at hand. With seven bodies meeting, though, negotiators appeared to be attempting a draining and complex balancing act of trade-offs between and across the agenda items. The one thing everyone agreed on was that several of the bodies needed to conclude their work for good. Even with the secretariat's careful planning, it was too complicated to field so many discussions at once. For the AWG-LCA and the AWG-KP, everyone agreed that Doha would be their final session.

Negotiators were stuck on defining the length of the Kyoto Protocol's second commitment period. The LDCs' preference for five years was shared by the G77 and other developing country groups. Concern that the protocol wouldn't cut enough emissions palpably increased when Japan, New Zealand, and Russia announced that they would join the United States and Canada in refusing to take on further reductions. All told, the Kyoto Protocol would now only cover about 15 percent of global greenhouse gas emissions.

Developing countries brought the need for a shorter commitment period to the table again and again, but the EU and the Umbrella Group supported a term of eight years. The EU highlighted that its internal legislation was already in force for 2013–2020. The African Group representative threw his hands up in exasperation, saying that perhaps ministers could make the necessary decisions. Perhaps *they* would prioritize the climate over existing policies and go for shorter commitment periods that would allow more frequent opportunities for review and enhancement.

Finance would go to high-ranking officials too, as those negotiations weren't going well either. The presidency appointed a pair of ministers from the Maldives and Switzerland to move the discussions forward. They focused on the period of next year, 2013, until 2020 to try and assuage the calls from developing countries—like the LDCs—for a road map that marked how funding would increase to meet the promised $100 billion per year by 2020. What was lacking at the high table were financial pledges to instill confidence that this would happen. Even with ministers facilitating, no firm commitments from developed countries were forthcoming.

At least I knew where to find the debates about the Kyoto Protocol and finance. The negotiations of the new agreement emerged and faded at a hectic rate. The discussions that did take place focused on work planning so that the process would deliver a new agreement by 2015. However, increasing efforts to bring down emissions before 2020 were at the forefront of people's remarks now that so many countries had walked away from the Kyoto Protocol. Groups of both developed and developing countries stressed that only strong political signals would keep this work on track and prove the world was serious about addressing the climate crisis.

The process, unwieldy at best, seemed to have an absence of political oversight. I marveled at the Qatari presidency's inaction—especially compared to the frantic convening the South Africans had done in Durban. In the presidency's consultation rooms, old men with glassy eyes talked leisurely about things the LDCs held as critical without conveying a sense of urgency or much interest in what was at stake. To avoid despair, I shifted my attention to the younger government officials at the back of the room who noisily fussed with espresso machines in an unpracticed way. They clearly weren't usually the ones tasked with bringing people coffee.

Wherever agreement couldn't be reached, agenda items were carried forward to the negotiations' second week. Higher-ranking officials then began sorting out the remaining political questions before all decisions were forwarded to the COP president for final adoption. This meant that the second Monday of the negotiations was often the most fashionable day of the COP. The best in national dress was expected to greet arriving ministers and ambassadors, so we got to see the height of fashion from all over the world.

I couldn't decide which was my favorite.

"All right, Bhutan is definitely a strong contender," I told Marika. We were in the LDC office with Thinley, the ever-smiling delegate from Bhutan, the Himalayan kingdom between India and China. Thinley was a wiz with a camera and liked showing off pictures of his comically photogenic family hiking and mountain biking.

"You're looking very sharp today, Thinley," I said.

"Thank you," he beamed. He wore a patterned, knee-length robe that looked something like a kimono but was folded over at the waist and had wide, white cuffs that came up nearly to the elbow. He completed the ensemble with knee-high socks and loafers. The LDC office door opened and made way for Pa Ousman and Bubu. They dazzled in long, flowing safari suits cut from amazing blocks of color that swept over their long-sleeved, knee-length tunics and wide pants.

Marika shook her head in consideration, "They're looking very smart as well."

"We're trying to decide who's best dressed," I told them and Thinley. Bubu obliged by spinning in a slow circle, which set the whole office laughing.

It was such a hard competition to call. The saris embroidered with gold that were worn by the women of Southeast Asia were a strong showing, as were the tropical flowers in the hair of the islanders who dressed in brightly patterned tops. Mbaye, the negotiator from Senegal, clothed his rail-thin physique in immaculate suits with boldly patterned silk ties that matched his pocket squares and socks.

I was also learning to identify the prevailing characters outside the LDC Group. Several of them had negotiated climate change for longer than I'd lived, and many had earned quite the reputation. Some of these reputations included nicknames—some fond, some not so fond.

"If Harry Potter says accounting rules one more time . . ." I overheard one delegate say to another in the hall. I snickered and kept walking. Despite the nicknames they gave each other, these negotiators were long-acquainted colleagues. Over the course of a day, they could both make ardent speeches reproaching one another around the negotiating table and laugh together over coffee during one of what would surely be a series of late nights.

Rather than meeting with others in long indabas, the LDC Group's key negotiators waited in the LDC office for word of what the presidency intended to do next. As we neared the scheduled conclusion of the COP, the waiting periods stretched from hours to days. At first, I filled the time between when the negotiations ended and when the decision's text appeared with conducting a very specific kind of conversation.

Given my new connections, I was determined to write about the successes and failures of technology development and transfer in the Least Developed Countries. My master's thesis was supposed to be on the technologies countries used to combat and adapt to climate change. Working in the negotiations was a privilege. Before coming to Doha, I compiled a list of people from all over the world that I should talk to. Conveniently, they would all be here.

I pulled Manjeet, the delegate from Nepal, aside after an evening coordination meeting. His home country had a comparatively high number of UN technology projects that I was eager to learn about. I couldn't tell if his hesitancy was about talking to me in general or if I was just another thing keeping him from dinner. When he begrudgingly agreed to an interview, the resignation he emanated almost made me laugh. Mostly it reminded me of how nice I thought Nepalis were when I visited. I got the sense that saying no wasn't culturally acceptable.

He knowledgeably answered my questions, nodding while he detailed the projects he had heard about and some of the problems the government ran into when implementing them. I followed along as best I could, but I had lots of questions—even about the questions I wanted to ask.

"So, the UNFCCC's database lists several projects to prevent GLOFs." I let my confusion show. "Can I ask you, what are GLOFs?"

I hoped he would welcome my honesty.

"Oh," he smiled. "GLOF means glacial lake outburst floods." He studied me for a moment. "You've been to Nepal, yes?"

"I have." Of the forty-eight LDCs, Bangladesh and Nepal remained the only places I'd visited. I remembered the startlingly high sky and white peaks guarding Kathmandu, the diversity of dark-haired faces, and the plates of delicious *momos*, a dumpling popular with locals and tourists alike. My stomach rumbled and I wished for the green ones, with the chili sauce.

"Good," Manjeet said, interrupting my craving. "Then you know the mountains surround us."

The staggering point of Mount Everest, which I had seen from the plane that carried me over the border into China several years ago, filled my mind. The Himalayan range was shocking, even to someone who grew up in the shadow of an active volcano. The mountains of home rise in a series of distinct peaks—isolated summits between which the snow melts and green hills roll. The Himalayas were a seemingly impenetrable wall of jagged height and ice. Reverential and immense, they run the entire length of a continent.

"The Himalayas are full of glaciers," Manjeet continued. "Climate change is melting the glaciers into bigger and bigger lakes of water. If the glaciers melt further, the ice dams holding the lakes in place could burst."

I tried to picture this. Gullies of ice carved into the rock, melting mineral blue, season after season, high up, above where people lived. The glacial ice would lock the water in place until winter returned to freeze it. If the ice that formed the dam gave way, everyone lived downstream.

"They're a real problem. You have no warning. Water rushes down and destroys whole villages," Manjeet said. "They are very dangerous."

"Is there anything you can do?"

"You can build warning systems so that people can get to safety. Or lower the lakes to decrease the risk. Or build dams to slow the water. It is difficult. The lakes are very remote, and these actions take resources."

"How many lakes are there?"

Manjeet laughed. "Too many to count. I don't know if anyone knows. You know what roads are like in Nepal."

Nepal's winding mountain roads were not for the fainthearted or those suffering vertigo. Their turns, which didn't have guardrails, gave you clear views of valley floors thousands of feet below. And the jolting, unpaved curves ran only so far. The end of a track meant going on foot over trails and suspension footbridges, packing in days' worth of gear. Locating glacial lakes over an entire, nation-spanning mountain range seemed an insurmountable task.

It wasn't just Manjeet's country either. The Himalayas span Nepal, Bhutan, Afghanistan, Pakistan, Myanmar, China, and India. Billions of people live downstream.

By after dinner on Thursday, I had conducted all my interviews. Marika had gone back to the hotel after the 7:00 PM LDC coordination meeting ended, and I wondered who would still be hanging around the LDC office waiting for news. When I opened the door, Pa Ousman and Achala were arranging rows of chairs along opposite ends of the room. This was not a good sign.

"Tonight's going to be a long one," Achala yawned. "You can go back to the hotel if you want to. We won't have new decision text until at least 10:00 PM."

I paused for a moment, counting the remaining number of chairs and my desire to work through the night. My internal debate didn't last very long. If I had learned anything over the past year, it was to sleep while you could. On my way out of the conference center, I passed Manjeet and Sandra—his counterpart from Togo—outside one of the few coffee stands that remained open. A handful of other delegates were camped out on nearby couches, comfy chairs being a valuable COP commodity.

When Manjeet smiled in recognition, I asked, "You're in for the long night too?"

"What else can we do?" he said with a shrug.

Sandra seemed less thrilled with the responsibility and turned her hands up in resignation. I wished them both good luck, then glanced at my watch and picked up my pace. The shuttles left on the half hour, so just missing one meant a disappointing wait in the sand-blown parking lot. If only I could make the climate negotiations move faster too.

———————

In the end, a political package of decisions was bunged together in the COP's final hours. Among the decisions was the pledge of thirty-seven developed countries to reduce their emissions under an eight-year second commitment period of the Kyoto Protocol. Though the protocol would survive until 2020 when the new agreement was expected to come online, the time frame essentially froze reductions targets until then. Finance decisions were largely postponed until the next session, their only tangible outcome the endorsement of the Republic of Korea as the host of the empty Green Climate Fund.

At the closing plenary, Minister Al-Attiyah gaveled decisions through, one after the other, without stopping for statements. "Hearing no objection, I decided!" he yelled into the microphone.

Even as a COP sophomore, I knew the presidency was meant to say, "Hearing no objections, it is so decided." They were also supposed to wait before gaveling to ensure that there were indeed no objections. Reactions to Minister Al-Attiyah's controversial mode of decision-making ranged from shocked laughter to whole delegations standing in dismay.

"I am not saying what is in store is a perfect package," Minister Al-Attiyah said after his gavel stopped swinging. "Perfection is just a concept. If great minds like Plato and Socrates were in the COP presidency, I assure that even they would not have been able to deliver a perfect package."

My mouth popped open at his arrogance. I looked across plenary to the fuss growing at the Russian Federation's chairs. The Qataris had completely ignored Russia's attempts to speak before the gavel flew, something they were clearly not going to let pass unmarked. When statements where finally allowed, Pa Ousman was quick to lament the number of elements the decisions left unaddressed.

"The package only promises that something *might* materialize in the future," the AOSIS chair said, picking up where the LDC's had left off.

Seated behind Pa Ousman, I gazed ahead, my brow crinkled in frustration. The negotiations had essentially done nothing since we left Durban except create a work plan. At this pace, agreeing to a new treaty in three years seemed absurd—let alone acting fast enough to address the climate emergency already happening in The Gambia and Nepal. I thought the UN would move things forward, but its cumbersome, multibodied approach was not geared to enable quick decisions, particularly under such leadership. I watched with dismay as Russia and Poland slammed their fists on the tables to punctuate their own disappointment. When statements finished, I prepared to leave with little more than sleep deprivation to show for the effort.

I could think of only one thing that had really changed.

Doha marked Pa Ousman's last negotiation as chair of the LDC Group. We gathered for a handover ceremony before the truly sleepless nights of the second week began. The chairmanship rotated every two years, shifting between the Least Developed Countries of Francophone Africa, Anglophone Africa, and those in Asia and the Pacific. Once appointed, the chair was the public face of the LDCs. That person's knowledge and negotiating skill would affect the bloc's power to inspire internal agreement and engage in external alliances. In line with the group's established rotation, the next chair was to come from an LDC in the Asia Pacific region, and members had selected Nepal to take the torch.

Marika booked a large room for the LDC Group complete with a standing podium and a cameraman at the front for the event. At the ceremony, long-standing members took the stage to praise the LDC Group and its outgoing chair. They talked about how far the group had come since forming in 2000, all the trials faced together, how the group had grown from strength to strength, and how the chair was no exception. Delegates raised their flags to applaud Pa Ousman for a job well done. From the top table, Bubu beamed while Pa Ousman looked down, overcome with emotion. He accepted the LDC Group's thanks for his service with relatively few words, after which he rose and shook hands with the Nepali delegation amid the frantic clicking of cameras.

Before the ceremony ended, Bubu took the microphone. "I can't forget to say thank you to Achala, Marika, and Biryani for all the hard work they have done to help the Gambian delegation."

I laughed from the front row. Pa Ousman poked him.

"Ay! I said it again. It's not biryani," Bubu laughed. "Thank you, Bri-hann-na! How could I forget?!"

Pa Ousman smiled and reached for the mic. "Yes, thank you to IIED, and thank you, colleagues. It has been a privilege to be your chair," he said in farewell.

The room applauded energetically, and we all rose in ovation. For a year now, I had watched Pa Ousman build alliances and push for faster action. I wondered how his leaving the chairmanship would affect the negotiations and my future in them. I smiled as Bubu waved at me, laughing again. I liked working with the pair of them and hoped my days as an honorary Gambian could continue.

"He's been the best chair we've had," I heard a negotiator say to his neighbor. Though I had no point of comparison, I agreed.

My last night in Qatar found me cross-legged on the balcony of an empty hotel room. I had just hugged Marika goodbye so that she could catch her flight to London. In direct contrast to the temperature awaiting me back in the States, the evening was remarkably pleasant for sitting beneath the sky. I watched the empty square and mosque in contemplation.

"Abba?" I called aloud.

"Yes, my love."

Thank You for the strange and wonderful journey that is my life, I continued in thought. Of all the things I imagined, being thanked by the Least Developed Countries at the UN climate negotiations in Doha, Qatar, wasn't one of them.

I smiled in disbelief.

How do You come up with this stuff?

Laughter rumbled through my head.

"I knew it would make you happy."

It does, Abba. So much. I like this work immensely and the position You have given me in it. I like being a researcher. It's hard. No, challenging, I corrected myself. *You have to reach out to people, and there are more questions than answers. But it all fits. I'm sorry I complain about it. There are drawbacks, but complaints don't adequately reflect the magnitude of my appreciation. Thank You,* I thought in earnest.

I smiled into the darkness, radiating contentment.

We were not done.

"My love?"

Knowing what He wanted, I attempted to completely empty my mind. Though previous experience had taught me that hiding—even in my own head—did not work, I was suddenly eager to have another go.

Time passed and the guilt grew. I really shouldn't ignore Him. Eventually, I answered, *Yes, Abba?*

Only it wasn't Him who needed to do the talking.

He waited.

Noelle's picking me up when I land, I hedged. *Thank You for her, my best friend . . .* my thoughts trailed off, avoidant. I closed my eyes.

The broken cycle of my relationship with Dad filled my mind. For months I had let the rims of my eyes run red from tears trying to riddle out what I should have done, what I should have said, what love should cost. Before I left for Doha, Mom told me that Dad blamed pain medication for his behavior. I didn't buy it. I was sure the medication they gave recovering lung transplant patients was powerful stuff; I just recognized his behavior as nothing new.

It was really only a matter of time, wasn't it, Abba? Nothing's changed. Yes, Dad has new lungs. But he's the same man—with the same daughter.

"Yes, my love."

I hope it's easier for Mom with me gone. In a way, it's strange to be treated like this again. To remember . . . I tensed, abruptly stopping this chain of thought. My eyelids squeezed tighter together involuntarily.

I exhaled, letting them release and take in the view.

Aloud in exaltation I said, "Thank You for not giving up on me."

7

HOPELESS

Kelso, Washington
January 2002

"You are no longer my daughter."

Dad's exact sentence played on a loop inside my thirteen-year-old head. After he said it, he ignored me. Left me blinking outside his car in Coweeman Middle School's parking lot.

Then there were no more rides, no more words. He didn't answer when I spoke to him. He didn't answer when I asked him why.

Rejection boiled to the surface, dissolving me into tears at the strangest times. When I cried in front of him, he didn't respond. I started wishing he would hit me again. At least than he would have to acknowledge my existence.

Mom started traveling for work. In her absence, mealtimes changed. Dad asked Chanteal what she wanted for dinner, told her when it was ready. They would sit at the table and I would drown, completely unsure. He would be there for her, and she would be there for him. They had a relationship that I did not. I learned that I didn't know how to cook and that I had taken hot meals for granted. Cereal suddenly became very important to me. I spent a lot of time at friends' houses, trying to escape.

Days turned to weeks.

There was no communication. I dreaded Mom saying, "Dad will take you." It would always go wrong, end with Mom telling me I wasn't turning up where and when I was supposed to. I didn't know where that was.

Alone, I walked narrow highway shoulders and across town in the dark, praying I would find Mom. Hoping that the people who offered me rides were nice ones.

I did try. If I could just change—if I could be better—things would be fine. But I didn't know how. I didn't know what he wanted. With each failed attempt, I believed him more. It was my fault. If I were less difficult, less defiant, more lovable. If I weren't so stupid, so fat, such a failure, he would take me back.

Weeks turned to months.

My body started to hurt. A pain grew in my stomach and I couldn't seem to remember how to laugh. Then how to smile. I enjoyed nothing, looked forward to nothing. Could see no future, no end. I. Existed.

Teachers started asking questions. Friends' parents too. Dread and depression hunched my shoulders, kept me awake at night.

I could not read him.

I could not understand.

It was hopeless. I was hopeless.

———————

I went there to die.

One January midnight, I stood alone on the edge of a cliff. I watched the fog filling the abyss below, its gentle swirl and the promise that it held, as pure and irrefutable as gravity. The rock quarry less than a mile from my parents' house had been remarkably easy to find through the winter woods—coatless and without a flashlight. What was the point of such things anyway, when my objective was to end life?

I welcomed the loss of feeling as a taste of things to come.

Nearly a year ago, Dad had told me that I was nothing. I believed him. He hadn't spoken directly to me ever since. Mom left. I didn't know when she was coming back.

An unwanted life. Painful. Pointless.

In death I would feel—nothing.

The sweet relief of nothing.

The cliffs. The quarry had blasted to get to new rock recently, carving sheer faces hundreds of feet into the earth. A swan dive would do it. If the fall didn't kill me, no one would find my body until the Monday two days from now. By then the cold or an animal would have surely finished me off.

Simple. Effective.

Finally, a way out.

I backed up from the edge and got ready to run. The loose rock I tipped over in doing so heralded my way. Down. Down. Down. Before the leap, I had only one thing left to do.

And that was when I first heard God.

"Go home."

The words reverberated against my eardrums. Alone on a cliff, audible to the fourteen-year-old come to die.

I had told the God of my education that if He was really real and if He didn't want me to do this, then He needed to show up. Otherwise, life was just too painful to continue. Empty, I had spoken my last resort to the rock and the mist. Empty, I had prepared to run, to leap just before the edge.

"Go. Home."

I fell to the ground. Faceup.

I lay there frozen like the earth while the words rebounded in my mind. The midnight fog enveloped me. I watched the white crowd out all vision, cloak me in a haze of indistinguishable direction, numb the pain.

I lay empty for a long time.

Until eventually, I got up and did as He said.

8

HOW IS THAT FAIR?

Bonn, Germany
June 2013

"A MONTH AGO, the global concentration of greenhouse gases broke the records of human history," Manjeet said.

I stared up at him from the front row of the LDC Group's prep meeting. I was now a graduate of Brown University and possessor of a master's degree, which "officially qualified" me to work here, at the United Nations. This was my life as a full-time, bill-paying consultant of IIED, who now had her flights and hotel paid for. And who got to watch the risk I took in borrowing all that money slowly draw down. I was here, making a difference and earning a paycheck at the same time! I still couldn't really believe it.

I worked to support Nepal, the country that took over the LDC chairmanship following The Gambia's handover in Doha. Even with the change of chair, Achala had retained her position as legal advisor and brought her team, Marika and me, along with her, here to Bonn, together, a week before the June climate negotiations. My fingers typed with eager attention.

"The level of carbon dioxide in the atmosphere has reached four hundred parts per million—the highest level for some three million years."

The PowerPoint slide behind Manjeet showed the greenhouse gas's potency zigzag up a chart at pace. Four hundred parts per million was well above the 350 parts per million threshold deemed necessary to keep our climate livable.

"This is truly alarming." He paused, letting the room of negotiators take this in. Manjeet was now a lead coordinator for mitigation agenda items. His presentation on the state of the climate was headlining the preparatory meeting.

"The year, 2013, marks the twenty-seventh consecutive year of above-average temperatures." He clicked through a series of world maps that showed a steady increase of red heat readings. "It already ranks among the ten warmest years since record keeping began in 1880, even though it's only June."

My notes were full of trauma, examples of abuse. Temperatures were off the charts in Australia, making 2013 its warmest year. Ever. Heat waves spiked readings of 120 degrees Fahrenheit and fueled mass heat stress die-offs—like the hundreds of thousands of bats found dead over Queensland. Southern China was currently gripped by an intense heat wave and drought too, with seven provinces receiving less than half their normal summer rainfall in 100-plus degree Fahrenheit weather. Millions of cropland acres dead.

California was enduring its driest year on record, with precipitation at just a third of average. The LDC of Angola and its neighbor to the south, Namibia—where one of every four people were chronically undernourished—were locked in a second consecutive year of extremely low rainfall in a string of thirty years that had tended toward dryness. And a drought in Brazil's northeast, thought to be the most severe in the last half century, continued from late 2012, with some areas receiving no rain for over a year.

"My home, Nepal, is receiving record rainfall." Manjeet's change in tone prompted me to stop typing and look at him. Increased global temperatures and the accompanying climate change amplified both dry and wet spells.

"This is the same in northwest India, which has received double its normal precipitation. The resulting floods and landslides have killed thousands of people." Manjeet stopped for a moment. Estimates of the region's loss of life stood at sixty-five hundred people, and the waters hadn't fully receded yet.

"I don't need to tell you how extremely vulnerable our countries are to climate change. These catastrophes are only going to get more frequent and more intense."

I knew this now in vivid detail. A few months ago, Professor Roberts had Brown's climate and development lab run numbers. The one billion people living in the LDCs were five times more likely to die from climate-related disasters than those living anywhere else in the world. From 2010 to this summer, stronger storms, longer droughts, and unprecedented floods exactly as Manjeet described had already killed hundreds of thousands of people.

"We are seeing these impacts everywhere," Manjeet said, pointing out the window. Less than a hundred feet from us, the Rhine had breached its banks, flooded with summer rain. Germany and central Europe were in the grip of flooding that had killed twenty-five people.

The climate crisis was not impending, something for future generations to worry about. Climate change was killing people now. Taking lives and livelihoods now. Everywhere. And things were only going to get worse. I heard it then, my shallow intake of breath. The feeling of climate change building in panic.

"Average global temperature is now about half a degree Celsius warmer than the twentieth century. As a group, we must continue to reiterate that temperature increase should be held to 1.5 rather than 2 degrees Celsius."

A two-degree temperature rise was the limit countries had promised to hold. Collectively they would keep their greenhouse gas emissions low enough to prevent warming the atmosphere more than this.

"We are already suffering. A two-degree world is not livable for us. We must insist on this, as a key priority for the LDCs," Manjeet ended, looking around the room. One by one delegates turned their country flags upward to speak.

"Tuvalu, please."

I turned in my seat, looking for the gray-haired delegate with glasses and a PARTY badge hung from a string of seashells. "Thank you, Manjeet," Ian began.

This was someone who could hold a room's attention. Ian taught at an Australian university most of the year. That's why Bubu called him "the professor," though he'd been involved in the UN climate negotiations even longer than Bubu had. It came across in the way Ian spoke. I coveted his ability to explain complicated things in a straightforward way.

"As you've so vividly demonstrated, limiting temperature rise to 1.5 degrees is crucial. I cannot overstate how paramount this is to Tuvalu. Climate change and the resulting sea level rise poses an existential threat to our islands and our way of life."

I remembered the glorious heat of the Pacific; the unrivaled ease of its people, the skirted men and the women with flowers in their hair; the impossible beauty of its beaches. Tuvalu's islands were halfway between Hawaii and Australia, cherished atolls in the middle of a vast sea. The country's highest point of elevation, at the time, was just fifteen feet above the ocean.

"Tuvalu's highest level of government endorses a 1.5-degree limit. The prime minister fully supports this position and has mandated me to implore the LDC Group to remain united in this fight. We must keep insisting that 1.5, and not 2 degrees, is what nations commit to, going forward."

After Ian finished, every other speaker joined in his support. From my seat between Bubu and Marika, I took in the room of LDC negotiators. They would enter the June climate negotiations united in this goal, resolved in the face of climate change's deadly reality.

They who stood to lose everything.

———————

"How is that fair?!" an Indian negotiator yelled. I sat behind Pa Ousman in one of Bonn's unremarkable meeting rooms listening to the negotiations of the new treaty, which people now called the "2015 agreement," thunder around the room.

"In conclusion," the delegate said, gathering himself, "India would like to underline that any discussion of the post-2015 structure must align with the convention's principle of common but differentiated responsibilities. We will not accept a dynamic interpretation of this, such as a two-step or hybrid process. We must remain focused on the agreed principles, which the 2015 agreement will also be subject to."

My fingers clicked across the keyboard, trying to keep up. I was pay-ing particularly close attention. The new LDC chair, Prakash, wanted a briefing. Prakash was a middle-aged Nepalese man who wore glasses and a no-nonsense expression. Relatively new to the UN climate negotiations, he requested far more detail in his briefings and talking points than Pa Ousman had. My current task was to write about national responsibili-ties under the 2015 agreement, summarizing the views of LDCs and other groups in a way that highlighted any differences in approach.

"Thank you, India," the facilitator said, looking down the list of speakers. "Brazil, you have the floor."

"Thank you," a Brazilian negotiator began. "I would like to align my statement with that of the previous speaker, the distinguished delegate from India. The 2015 agreement will set us on a new path, but it will not leave behind the foundation on which it is anchored . . ."

In Durban countries decided that, under the next international cli-mate treaty, all nations would reduce greenhouse gas emissions. It was this deal that held the negotiations together two years ago. In Doha parties decided they would assemble a draft of the 2015 agreement's text by the COP in 2014. We had a year and a half to put together a treaty that would be radically different from any existing climate agreement.

"It is with this context in mind," he continued, "that I would like to reiterate the Brazilian proposal, which addresses historical responsibility not just in terms of emissions, but also in terms of relative historical contributions to global temperature increase."

I carefully wrote down each country's proposal for defining reduc-tion targets, the time frame for doing so, and how to ensure enough was done. Pa Ousman shuffled the notes in front of him. The Gambia's flag had been up for over a half hour, so it was nearly time for him to speak.

After his chairmanship, the LDC Group sent a letter to The Gam-bia's president recommending Pa Ousman's appointment as a special climate envoy. The title was a first. No other LDC had an ambassador devoted to climate change, though most developed countries and an increasing number of developing countries did. Pa Ousman now ranked with the other diplomats present at the negotiations and could represent the group at a higher political level than before.

I was happy he was still negotiating. It meant remaining an honorary Gambian and working with the delegation I had come to know well over the past eighteen months.

"The scientific community is best placed to carry out the methodological work," the Brazilian said, concluding. "Establishing an expert group to discern developed countries' contributions to the temperature increase would aid in this regard. I'll end my remarks there. Thank you."

"Thank you, Brazil," the facilitator said. "The Gambia, you have the floor."

Pa Ousman clicked the button at the base of his tabletop microphone. It glowed red.

"Good morning colleagues," he began. "The Gambia is speaking on behalf of the Least Developed Countries. The LDCs see merit in elaborating an equity reference framework under the 2015 agreement. This framework would support the use of metric and nonmetric criteria, including historic responsibility, future needs, and vulnerability. This would fairly determine appropriate reduction targets, particularly for those who have contributed the least to climate change."

These notes I had on file.

I tracked what the LDCs said most carefully, sitting up straight whenever one of the forty-eight country flags went up. Bangladesh had already stated that reduction targets under the 2015 agreement must be clear, enforceable, and scientifically sound. Ethiopia proposed a hybrid approach to determine the targets, which was based on historical and per capita emissions, the global temperature goal, apportioned atmospheric space, and quantified emission rights.

There was so much to keep up with. Defining in brief all the group's positions about what should be in the new treaty was a lot of work. And though I knew what most of these concepts meant, some had elements I didn't understand. Trying to put them together was starting to make my head hurt.

Talk to Ian, I typed. Getting Ian to break this down for me would be by far the easiest path to comprehension. I just had to convince him, which might be a problem. Along with Bubu, Ian was taking a lot of flak this summer for eating multiple ice creams per sitting. This flak came mostly from me.

When the 2015 agreement negotiations had reached an interval the day before, I'd called Marika to coordinate a shared lunch break. This coincided with Bubu's, igniting familiar dietary arguments. During our usual dustup over the climate impacts of beef, Ian pulled up a chair at our cafeteria table. He had two Magnum ice creams on his tray.

"Hungry?" I teased.

Ian chuckled, unbothered. "Didn't your mother tell you to eat your ice cream?"

A great lover of chocolate, sugar, and full-fat cream, Bubu was quick to come to Ian's defense, "We can't eat too much ice cream." He left the statement hang there, begging a response.

"Why not?" Marika caved.

"We fly here over the Sahara to get here. It's so hot that it burns the ice cream away."

Marika laughed, I looked at Bubu like he'd lost his mind, and Ian's only response was to nod seriously, as though what Bubu had said made any sense.

"You two are ridiculous."

Ian unwrapped his first bar and smiled, "I'm just fueling up for the Sahara."

———

I pulled Ian aside that afternoon and asked him to talk me through the varying positions on the 2015 agreement. "Prakash wants a briefing," I said in explanation, hoping my ribbing earlier wouldn't put him off. He just smiled. He seemed happy to fill the gaps in my understanding, explaining what each concept meant and how they all might manifest in the new treaty. I asked for clarification until I was positive I could write the messages clearly, surged with relief when we reached the end of my questions. Ian knew his stuff. I liked that we were on joking terms now.

"How often do you go to Tuvalu?" I asked conversationally when the explanation portion of the discussion finished.

"I visit at least once a year," Ian answered.

"I tried to go once, when I was in Fiji," I said. "They told me the ship left last week and there would be another—sometime next month."

The philosophy of time that lived in the Pacific baffled me. And not just because several people tried explaining it to me in Fijian—something I had not anticipated but appreciated. I was used to being mistaken for a local in the Americas. I hadn't thought it would happen anywhere else. But then, I was also big and dark brown with curls of thick, black hair.

Ian laughed. "Sounds about right. How long were you in Fiji?"

"About a month. I was there for Christmas and New Year's five years ago." Man, that was wonderful. I spent weeks wandering, wet, in the impossible beauty, dipping from pool to beach and awed that atolls could gather an ocean too warm to cool yourself in, so warm it bleached the coral white as snow. Inhabiting a strip of land you could walk around in a single day confounded me.

I liked the hallelujahs of the Christmas service. I loved the fish buried to cook and served in coconut milk curry. I hoped I would never drink kava again. "A depression went through toward the end of my trip, so I thought I'd be there much longer."

They called it a depression. It wasn't a term I associated with weather, but I'd guessed a storm was coming. For days the sky thundered with rain. The wind whipped the ocean into giant, angry swells that hit the coastline and all who dared remain. The town I was staying in closed— the entire town. High water submerged the one highway. My joy at the sun-filled island evaporated as its inherent vulnerability became shockingly and inescapably clear. There were no alternate routes to anywhere. The only airport stopped answering the phone. Planes couldn't take off in the gales anyway. We were trapped, and the water was rising.

They started reporting drownings on the news. Eleven lost last count, three of the dead teenagers. From the last dry footholds, I could see the waters where they died.

Changing our climate meant gambling with the lives of everyone living on the shore, particularly those who had no higher ground to go to.

———————

The door to my very own Bonn hotel room clicked shut on the final day of the June negotiations. It was a beautiful summer evening; the church bells rang in a red fading of the light.

"Abba?" I called aloud.

God. My Father. Over a decade of intervention built this relation-ship—piece by piece. As the girl on the cliff, I believed myself nothing of value. For years following, I regarded any expression of my worth with suspicion, any measure of closeness with resignation that it would soon disappear. Broken and alone, I was first to cast myself down.

Only I wasn't alone.

The preaching I heard in my late teens transformed my image of God. It had me reexamining all verses memorized in childhood, reread-ing the Bible in its entirety. And I couldn't believe that either. I caught myself writing lines on bits of paper. *How precious God's thoughts about me, they outnumber the grains of sand.* Marveling. *You are fearfully and wonderfully made.* Mesmerized. *You are precious to me. You are honored, and I love you.*

He said such ridiculous things. They couldn't possibly apply to me.

The preaching kept me coming back, listening to sermons in stu-dios where I was meant to study architecture and while walking around campus. They talked about experiencing the love of God poured out as a *friend. One that sticks closer than a brother.* Surely not, only—I knew a God like that. The preachers talked about Him like they heard Him too.

For years I tried at all costs not to think about that darkest of moments. I certainly didn't talk about it.

But perhaps my experience of God was not as rare or delusional as I imagined. And if His Spirit was a friend, maybe I could ask Him. *Do not be afraid, for I am with you. I will strengthen you and hold you up.* I learned to hear that same voice, no longer audible but distinct. And I couldn't stop talking to Him, caught up in a new understanding of who God could be, reaching out and building up new narratives in my mind.

Coaxed into responding, I laughed at His jokes. I held His hand, even until the literal *end of the world*—where we traveled together. I trusted Him as I could not trust my own father. The first time I whispered, "Abba," I burst into tears. The ability to think of anyone as "Daddy" broke countless barriers constructed for my own protection.

I didn't know anything like this Love. Patience on an unhuman timescale. Kindness. Undemanding, eternally hopeful. *His faithful love*

endures forever. It was not inconsistent or determined by my perfor-
mance. Year after year phrases like *I have made you glorious* replaced the
others stuck in my mind. I still didn't always believe them, but eventually
I started thanking Him for making me *me*. An unthinkable opinion in
my adolescence: value in myself, just as I was, something I would not
have been able to hear before. I came to love Him back, enough that
daily conversations became integral.

So now, the God whose words I first heard on a cliff answered in
my mind, *"My love."*

The Indian delegate's words had chased me around Bonn for the
past week, as had the concluding sentences of the briefing I typed for
Prakash—the briefing he liked so much he quoted it at today's closing
plenary.

It's not fair, Abba.

I crossed the hotel room to stand at the window. I meant to stare
out at the Bonn evening in contemplation, but my own petulant reflection made me giggle, covering the depth of what I felt.

"What isn't?" He asked, undeterred.

Climate change, I thought. The people I worked for had polluted
the least, suffered the most, and lacked the resources to deal with the
consequences of the crisis. The forty-eight Least Developed Countries
had contributed less than 1 percent to the world's cumulative greenhouse
gas emissions. Less than 1 percent. On average, the billion people living
in these countries emitted 0.3 metric tons of carbon dioxide per year.
The average American, meanwhile, emitted 16 metric tons of carbon
dioxide per year. So those who polluted the most suffered the least and
used their resources to keep the worst impacts at bay.

Climate change was such an unjust mess.

I had heard Pa Ousman and Bubu talk about climate impacts in The
Gambia many times now, namely the rising water and the encroaching
salinity that was displacing millions. Also from West Africa, Mbaye in
Senegal and Sandra's home of Togo saw comparable impacts. Manjeet
and Thinley, whose homes were in the Himalayas, talked about the
rapidly melting roof of the world. Evans from Malawi saw drought and
seasonal rain variability spur migration, conflict, and famine in East
Africa. Even a small increase in global temperature would undermine

their agriculture, causing starvation. Sea level rise would kill those living and depending on the coast.

I thought that, given their position, the LDCs would have every right to take a back seat when it came to emissions reductions in the 2015 agreement. But as Pa Ousman had said during his chairmanship and Parkash echoed today as the Bonn negotiations ended, leadership was not something they would turn away from.

Their people experienced climate impacts every day. Their people were dying every day. The climate crisis was not something they would ignore.

The LDC Group's position was designed to ensure that *every* country took on as ambitious a reduction target as possible, even though this meant that they too would need to make drastic changes. Their foremost interest was a climate treaty that cut emissions most quickly. That mitigated the emergency the fastest.

I couldn't admire them more.

How can they do it, Abba? I thought, *lead, when the burden shouldn't belong to them, while we who are responsible do so little.* Sadness and anger saturated my mind.

"Given what they are facing, how could they not?"

It's not fair, I repeated.

As I wrote in Prakash's briefing, currently the United States favored an agreement wherein all nations designed their own commitments based on national circumstances, capacity, and other factors they considered relevant. I thought this approach sounded perfectly reasonable—for the Least Developed Countries. Their extreme poverty may well hinder their ability to reduce emissions, while at the same time developing their economies. This was hardly true for the world's foremost economy. If the United States didn't have the capacity to cut emissions, no country did. Besides, wouldn't such an approach inspire a race to the bottom?

The amount of effort we required to put us on a sustainable path was unprecedented, an energy revolution needed in half the amount of time as the Industrial Revolution, to service seven times as many people. To call that merely ambitious was a tremendous understatement, bordering on untrue. If we let countries decide for themselves, how would the UN ensure that enough was done?

I thought it a real shame a climate justice police force didn't exist. It would make things so much easier if scientists could simply break down the emission reduction requirements into nation-sized chunks and hand them over to the CJP for enforcement. If the Kyoto Protocol taught us anything, though, it was that assigning reduction amounts was a surefire way to see nations renege on their responsibilities. I remembered the Doha negotiations where one by one Japan, New Zealand, and Russia joined the United States and Canada in refusing to undertake their commitments.

The 2015 agreement held our hope for igniting a truly global effort. Politically, the idea of the UN or any international body telling the United States, China, or India what to do wouldn't work. Still, if a global body couldn't assign emission reduction targets, what, then, would be a fair approach?

What the hell is fairness anyway? I vented.

And in my mind, I saw Dad, smiling at my graduation from Brown last month when he posed for pictures. Mom told me the week before that he had bought a ticket to Providence. He gave me a card with a perfectly pleasant *Congratulations* written inside. Through no shared resolution, he was speaking to me again. Mom and Grandma told me to be grateful that he had decided to attend, reminded me that he had been through so much. I stood there, all my emotions undone. Unable to comprehend, my mind filled with questions that—surrounded by friends and family—I couldn't bring myself to ask. I was desperate to pretend too.

Now my hands covered my eyes, and I stood holding my face.

Living with the climate crisis was so like living with him: The stress and the fear. The constant risk of death. The building tension. The pressure and despair that impacts everything, underlies everything. The anxiety that kept me up at night. Violence was inevitable and would be inflicted on those least responsible. It was only a matter of time. How every day I watched those with power undervalue things that were precious, irreplaceable, treat them as worthless. And the silence around it, the isolation and the pretending, when it was not safe. Not wanting to voice such terrible things because talking about them would mean owning up to reality and the part I played in its perpetuation.

I don't have answers, I thought to Him. *To any of it.*

9

THE TRAVELING SHOW

Warsaw, Poland
November 2013

AT SOME POINT DURING THE AUTUMN OF 2013, I became a circus performer. Or at least that was how it felt, like I took a job for the traveling show and never settled down again. After the June negotiations, the bags I had packed in Providence following my graduation stayed packed. I just kept moving, logged on wherever I found myself in the morning to juggle drafting speeches, briefings, and talking points.

"You're sure you know where everything is?" Noelle asked, walking toward the door of her Kelso townhouse. She reached down one last time to scratch the golden retriever following her, then stood to hug me.

"Yup. I have your keys. Belle's leash is on the hook, and I'll text you if I need anything."

I waved at Kevin through the open door. He had been waiting in the car for the last five minutes. It was a morning in early July, the sun was coming through the trees of their corner lot, and they were off to California on an anniversary road trip.

Once their car turned onto the main road, Belle used her three legs to hobble over and press herself against me. When Noelle told me they had adopted a dog that had a leg amputated after chasing down a logging truck, I thought she'd be wild. Belle emoted happiness in a

141

way that was absolutely endearing. As she turned her head up to look at me, a bright pink tongue lolled out the side of her upturned mouth. Ridiculously cute.

I lived my weeks in Kelso largely from Noelle and Kevin's. Chanteal had moved to Buffalo, New York, to get a graduate degree in library science. Mom and Dad were at home, and as Dad was apparently speaking to me again, I got to store stuff in my childhood bedroom free of charge. Otherwise, I kept my distance from the house. The whole situation just made me angry. Why should he, king of mood swings, dictate so much of my experience?

Not voicing this kind of question and maintaining the peace meant some time with Mom and ensured that my money went toward repaying student loans rather than renting, for storage or otherwise. I tried to convince myself that I wasn't yet another advanced degree holder who lived with her parents. The truth was, I didn't know exactly *where* I lived. Consulting for IIED meant I possessed a steady income regardless of where I worked. I could finance going wherever I wished, so this seemed the perfect time to visit the friends I'd missed while away at Brown.

In August I stayed with Michelle in Boise. "So, how's life in the double-wide?" I asked her and her sister, Andrea, as their car pulled out of the airport parking lot. Because she was their chief resident, the VA provided Andrea with accommodation that Michelle, a medical student on rotation, was happily sharing rent free. I imagined that in most cities accommodation for doctors generally meant a flashy apartment or house. In Boise the provided accommodation was a double-wide trailer, one I had visited three times in the past eight months, a perk I wasn't going to let either of them forget.

"Now, because we're fitness Barbies these days, we're running tonight and hiking tomorrow," Andrea instructed. The three of us were amid a renewed commitment to getting in shape. "I assume you two are going to have some kind of dance-off as soon as we get home?"

"Who needs to wait until we get home?" Michelle hit the volume on her iPhone, which traveled through the auxiliary cable and out the speakers, and she and I morphed into undiscovered music sensations, lip-synching and chest popping until the car stopped.

"I'm going back to the hospital," Andrea shouted above the music. "Get some work done and be ready to run at 6:00 PM!" Michelle and I tumbled out, singing at full volume.

From Boise, I headed to the Midwest for a special birthday. Grandma was celebrating her eightieth in style, and Mom, Chanteal, and I joined her in Ohio for the occasion. She radiated savvy from her hat to her heels walking into the church hall my aunt had decorated to the hilt. The extended family had gathered. The Thorbahns, the Fizers, all the cousins and friends raised their glasses and cupcakes to her accomplishment, while my sister snapped photos and Grandma smiled wide.

Then it was off to London toward the end of September for a series of meetings at IIED with Pa Ousman and the Britain-based crew. Talking shop over tea in the offices I had heard so much about extended my sense of team beyond the negotiations, but it was seeing Marika in her natural environment that I best remembered. She invited me 'round to meet her partner, Tom, and have Mexican food—our strongest common interest. It turned out that Tom's love of my favorite cuisine was so great that he elected to make his own tacos and hot sauce from scratch. Over too many servings, I learned that he also shared my obsessions with peanut butter and Beyoncé.

"I think this is the start of a beautiful friendship," I told them both as I left their flat, waving goodnight and heading back to my Airbnb.

By October Erina and I were catching up via walks along New York City's High Line and over plates of Greek food in Astoria. Erina had moved from Paris to New York City for the second year of her master's in international relations. We had seen each other regularly during my last year at Brown, as Providence and New York weren't all that far apart.

Afterward I took the bus north to Rhode Island to hang with Becca in the same cafés where we'd spent all those hours writing our theses just months ago, thrilled to relive the ease of simply being together in the restaurants and shops where we had spent the last two years nearly inseparable. She had a job in renewable energy now.

Then one November morning, I stood on a Polish street corner trying to get my bearings in the cold. The city of Warsaw had the grayness of a place that endured long, bleak winters. A parked bus across the street had a blinking neon sign in one of its windows that flashed OPEN. When I got closer, I realized that it had been converted into a bar. The murmuring of people chatting in an unfamiliar language became the muffled soundtrack that followed me while I matched street signs to the names mapped on the paper I carried. Until, suddenly and unignorably, a single word stood out among the din.

"Bri-hann-na!"

I laughed unbelieving. Bubu was standing beside me on the sidewalk, having seen me while making his way back to the hotel.

"Bubu!" I exclaimed, surprised at the bizarreness of journeying across the world only to end up seeing the same faces.

The next day as I made my way to the LDC Group's preparatory meeting, I ran into Hafij from Bangladesh. I had always thought that Hafij's easy smile contrasted oddly with the dark circles under his eyes. Today he joked, "It's amazing. I've never been to eastern Europe and yet I know so many people in Warsaw."

We walked together for a while, waving to the negotiators he recognized on the street as we made our way toward the LDC Group's preparatory meeting. The venue was the Palace of Culture and Science, a gargantuan building gifted to Poland by the Soviet Union in the 1950s. Approaching it along the six-lane roads only added to the dwarfed feeling conjured up by the Soviet-era cityscape.

Inside, up an elevator, and down several halls, the windowless conference room looked very much like the first UN meeting room I'd entered in Durban. The secretariat must move most of the equipment we used from place to place, because the rooms were nearly indistinguishable, one from the next. We sat at the same white, angular tables topped with black tabletop microphones and white flags. We were the performers, the secretariat moved our big top, and every few months, we convened again to dance the next show of the season. I blinked at this realization.

Afterward, I surveyed the group, looking for familiar faces and finding Marika's first. I waved at Sandra and Manjeet, who were busy compiling a coordinators list, still hardworking as ever. Mbaye stopped to give me a confusing number of cheek kisses. Though I understood that a kiss on each cheek was expected, I was perfectly happy to stop after one. Mbaye confounded me by leaning in for a third and then laughed at the expression on my face before moving off to greet others. I wondered if three was the standard in Senegal or if Mbaye just liked to baffle people. From the very front, Achala waved Marika and me forward.

"Good morning," we said in unison to Achala and Prakash, who were already congregating near the top table to kick off Prakash's first COP as LDC chair.

"*Dzień dobry*," Prakash smiled.

I laughed, impressed. "Is that 'good morning' in Polish?"

He nodded with a beam.

I learned in Bonn that Prakash liked to pick up phrases of the local language wherever he traveled. He'd carried around a small German dictionary during our time together in June.

"*Dgin dobry*," Marika responded, trying her hand at it, while another Nepalese man in a smart suit walked over to where we stood.

"This is Ambassador Bhattarai," Prakash said introducing Marika and me.

"Nice to meet you," he smiled, shaking our hands in turn. "I am Nepal's permanent representative to the United Nations in New York. Thank you for all the work you do with Achala to help the LDC Group."

He had a smooth, unpolished accent and looked me in the eyes as he spoke. "Prakash, shall we call the meeting to order?" he said, moving toward the top table.

I watched them go, a bit confused. The chairmanship position was tied to a single person, yet Nepal seemed to abide by this in name only. As the official LDC chair, Prakash opened the discussions. Afterward, he handed over to Ambassador Bhattarai, and several negotiators also tilted their heads to one side, trying—like me, I assumed—to work out who was in charge. Prakash and Ambassador Bhattarai remained at the top table throughout the LDC Group's preparatory meetings. This created a slightly awkward dynamic when they both tried to defer to the other whenever a decision needed making. New to the UN's climate

negotiations, Ambassador Bhattarai joked that the barrage of indecipher-
able acronyms had him constantly asking questions.

Remove as many acronyms as possible from the speeches and talk-
ing points, I noted, not quite sure now who would read them. I should
prioritize getting to know the ambassador better as well; it looked like
he would need the support.

"Ready to go?" I asked Marika at the close of the second day. I waved to
Achala, Prakash, and Ambassador Bhattarai as they headed out the door
of the nearly empty room.

"Let me just write down what Prakash asked before I forget," she
said, rummaging in her bags for a pen. As keeper of the chair's schedule, Marika was always the one given last-minute tasks. I had already
switched my computer off and was looking forward to an evening out.

When she finished her note, Marika continued to prove an excellent
travel buddy. We had compared lists of things to see in Warsaw when
we arrived and were making steady progress. Tonight's outing was to
the well-reviewed Jewish History Museum.

We walked the quiet blocks between our hotel and the museum
under the cover of her umbrella, emerging from the drizzle to stand
in line with the handful of other tourists queuing for tickets. Inside, I
couldn't get over the before and after photos of Warsaw's Old Town.

"Marika, come look at this," I said, waving. The large photo on my
left featured a bustling city center of festively decorated buildings around
a square. In the photo to my right, the square remained but was entirely
surrounded by abutting piles of rubble. The two looked like they were
in reverse order—after then before.

"The Old Town was the center of the resistance during the failed
Warsaw Uprising of World War II," I read aloud. "The German army
individually demolished each building. Many of the demolitions were
filmed and aired to the occupied public. By 1944, almost 90 percent of
the Old Town had been destroyed."

We stood there exhaling, trying to take it in. After the war ended,
it took the Poles nearly fifteen years to reconstruct the Old Town brick

by brick. We drifted off to finish reading the bleak and well-presented history. In the days that followed, we took to exploring the meticulously reconstructed historic center, hunting for vegetarian Polish fare and sampling flights of vodka together.

The COP itself was held in the national football stadium, a fifteen-minute tram ride across the river from our hotel. Figuring out which tram to take was made remarkably easy after day one, as many of the negotiators wore a kind of UNFCCC uniform. Traditionally, each host nation gave a small gift to welcome negotiators to their country. In Durban, the South Africans handed out beaded key chains shaped like safari animals. In Doha, the Qataris gave us notebooks in a canvas bag. On the opening day of Warsaw, the Polish gifted light gray hats, scarves, and mittens all emblazed with the COP 19 logo.

I laughed at first, yet as the November weeks stretched on and threatened snow, they came in surprisingly handy, especially to the delegates from places where you would never need such things. Most mornings, I followed a gray hat or two down to the stadium.

"So, Bubu, what needs to happen here in Warsaw?" I asked as we rode in together on day two. Stepping off the tram, I matched my strides to his as we followed the stream of delegates heading for the long pedestrian walkways and up the flights of external stairs that led to the stadium. Like Ian, Bubu's tenure at the negotiations was vast. He knew just about everything. Getting his take on what to look for formed an important part of my ability to understand the process.

"It's finance and the mitigation commitments that are the most important," he huffed. "Ay! Why do they build so many stairs?"

I reached over and took the briefcase he carried, to lighten his load. I had grown very fond of Bubu's uncle-like presence. The familiarity had solidified for me last year, when Bubu inadvertently became integral to my strategy for dealing with sexual harassment. Overhearing an unwelcome advance at an LDC meeting, Bubu crossed the room to tell the delegate off. He, after all, was the eldest member of the Gambian delegation and he did not appreciate people interrupting my work with lewd innuendo. Had the man no shame?! Bubu's confrontation worked far better than anything I had come up with. From then on, I handed him the unwanted business cards and phone numbers written on slips of paper—ideally with their owners close enough to see me point them out. They tended not to bother me after that.

We came to a halt to one side of a wide landing, letting people go by. A few stopped to say "Good morning" in the November chill. When one wiped his brow with a handkerchief, I realized that the small talk was an excuse to break the climb for them too.

After several minutes, Bubu seemed to have caught his breath. His face told me that he wouldn't mind stalling a bit longer, so I picked up where our conversation left off.

"Climate finance means a road map to the $100 billion, right? And reduction commitments under the 2015 agreement?" I asked, verifying. The LDCs wanted developed countries to get serious about the promise they'd made in 2009. The group demanded they map out the annual increase of public finance contributions needed between now and 2020—by which point developed countries had promised the total would reach $100 billion per year.

This COP was also the moment for nations to decide how they would frame their emission reductions commitments under the 2015 agreement, following their negotiations in Bonn. Countries would need the two years between now and 2015 to prepare them so that they could be submitted together with the new treaty. This I had gleaned from all my briefing writing.

"Yes," Bubu answered. "Developed country commitments are easy. We have the model of the Kyoto Protocol. They must take quantified economy-wide emission reductions targets."

Under that protocol, developed countries accounted for reducing greenhouse gas emissions generated across all sectors of their economies. To do this, they needed data to generate baselines and accurately measure emissions outputs over time.

"For LDCs, this is not possible. For some developing countries, it is," Bubu said.

"So, we have to agree on what the commitments should look like," I summarized, slowly turning toward the stairs. The cold was getting to me.

"Ay!" Bubu cried as he started to climb. I couldn't tell if this was meant in agreement or trepidation or both.

Once Bubu finished his ascent, we lined up to clear security, where afterward I walked through the halls of the conference center with a sense of expectation. I understood the UN climate negotiations infinitely better than I had two years ago—the labyrinth was decipherable now. In preparation, I had read the agendas of the bodies at work and looked up the meanings of any items I didn't recognize, pleased that a glance over most of them gave me an idea of what they would discuss. I could find just about anything—given enough time—on the UNFCCC's convoluted website. I recognized several faces crossing the corridors and happily took the time to "Good morning" the national delegates and former-interns-turned-secretariat-staffers I passed in the halls.

Seated in the plenary, I fidgeted behind Bubu in the second row of The Gambia's four chairs. The rest of the hall waited with us. We were due to meet the two people who would lead the UN climate negotiations' most important body for the foreseeable future. They would take the reins of the negotiating body responsible for delivering the 2015 agreement. Developed and developing countries had tussled over this appointment for months and only recently agreed that, given this work's importance, both groups would be represented in the chairmanship. I looked up at the top table as the Polish COP president finally welcomed to the stage the newly appointed cochairs, two middle-aged men named Kishan and Artur.

Kishan, the one with wavy, dark hair, greeted us in a faintly Caribbean accent. The agenda I had open told me he was from Trinidad and Tobago. Next to him, Artur wore speckled glasses and a checkered suit. The country listed against his name was the European Union. Kishan was saying, "By the end of the Warsaw session, half our allotted time to negotiate the new agreement will have passed."

I marveled at this truth.

"We need to shift gears in order to move forward, rather than reverse," Kishan went on. First up was to define the content and elements of the new agreement, including commitments to it. "Artur and I will facilitate consultations that will take place in plenary. Hopefully, this formal setting will provide for a dynamic, transparent, and inclusive exchange."

After the last country flag went down, the pair rose in unison to leave. Heads turned to watch them depart. I chuckled to myself. These two were the people everyone now wanted to know.

———————

In the hubbub of the afternoon, Hafij filled me in on what he was tackling in Warsaw. We were waiting in the LDC's allotted room for the Tuesday afternoon coordination meeting to start. Hafij's home was the only other LDC I had visited. Bangladesh held a population of 158 million people— approximately half that of the United States—yet its territory was about the size of New York State. This made Bangladesh one of the world's most densely populated countries, something that was not lost on me during my time there. India's neighbor to the east, Bangladesh occupies the world's largest river delta, a low-lying plain of green Sundarbans mangroves spiderwebbed with tributaries that flow into the Bay of Bengal.

In a switch from following the topic of adaptation, Hafij had recently taken up the issue of loss and damage—the irreversible economic and noneconomic casualties of the climate crisis, a reality made vividly clear to everyone at the negotiations yesterday. At the opening plenary, a negotiator I didn't recognize took the floor. His name was Naderev, and he was from the Philippines. He had straight dark hair that fell across his forehead, and he wore glasses. Busy writing something when he began, I was only half paying attention.

I stopped typing to listen undistracted once I heard the tenor of his voice.

He was crying.

I had never heard anyone cry in plenary before.

"Super Typhoon Haiyan made landfall in my family's hometown, and the devastation is staggering," he said. He was reading a prepared statement, but he couldn't go on. He choked up and looked down, away from the camera that was projecting his image on giant screens around the room. With a surge of sympathy, I watched his mouth crush into a tight line until he recovered himself enough to continue.

"I struggle to find words even for the images that we see on the news coverage, and I struggle to find words to describe how I feel about the losses. Up to this hour, I have agonized waiting for word on the fate of my very own relatives."

Google brought up apocalyptic images on my laptop. Super Typhoon Haiyan was the strongest tropical storm to ever make landfall. Winds had reached 235 miles per hour. An estimated eight thousand were dead. Millions were homeless. Two-thirds of the country lay in ruins.

The Philippines was used to cyclones. The headlines read that despite the government's massive efforts to prepare, the traditional practices had not worked. The storm warnings had sent people to shelters that had weathered countless cyclones, shelters built to keep them safe. Super Typhoon Haiyan was more powerful than anything they could have anticipated. It was more powerful than any cyclone that had ever been. The people died in the shelters. Nothing could have kept them safe.

The pictures were unthinkable. A vast wasteland of mud, debris, and dead bodies. Gone. All gone. The floor of my stomach dropped, and I struggled to hold back tears.

"What gives me renewed strength and great relief is that my own brother has communicated to us that he has survived the onslaught," Naderev continued. "In the last two days, he has been gathering bodies with his own hands. He is very hungry and weary, as food supplies find it difficult to arrive in the hardest hit areas."

Naderev paused again to gather himself; then he pressed on.

"Mr. President, during these last two days, there are moments when I feel I should rather join the climate advocates to peacefully confront those historically responsible for the current state of our climate. The selfless people who fight coal, who expose themselves to freezing temperatures or block pipelines. In fact, we are seeing increasing frustration and thus more increased civil disobedience."

Protesters had lined the ring road surrounding the national stadium that morning. The Polish prime minister had recently announced plans to build several new coal power plants, all while preparing to host the UN climate change conference, a hypocrisy many would not let go unmarked.

"The next two weeks, these people and many around the world again remind us of our enormous responsibility. To the youth here who constantly remind us that their future is in peril. The climate heroes who risk their lives and reputations to stop drilling in polar regions and those communities standing up for sustainable sources of energy, we stand with them. We cannot solve problems with the same actions that created them!

"Mr. President, I express this with all sincerity." He paused. "In solidarity with my countrymen who are struggling to find food back home and with my brother, who has not had food for the last three days, in all due respect, Mr. President, and I mean no disrespect for your kind hospitality, I will now commence a voluntary fasting for the climate. This means I will voluntarily refrain from eating food during this COP until a meaningful outcome is in sight."

I had never heard anything like this at the UN climate negotiations.

"Until concrete pledges have been made to ensure mobilization of resources to the Green Climate Fund. We cannot afford a fourth COP with an empty GCF," Naderev built in emphasis. "Until the promise of the operationalization of a loss and damage mechanism has been fulfilled."

I gaped. And I wasn't the only one. Everyone had turned to face him.

"We can fix this!" he proclaimed.

Tears welling in my eyes, I realized that this was exactly what I wanted. What we needed.

"We can stop this madness," he ended. Then he broke down and wept openly into a red and white handkerchief. The quiet room stood and applauded. Many, like me, wiped away their own tears, the negotiations' purpose never made so painfully clear.

We had to get this done. Now.

"We tried to discuss loss and damage in Doha. The islanders too," Hafij told me the next day, continuing our discussion in the empty meeting room.

"Oh, yes. I remember you saying last year," I said. "There was a fight, right? Developed countries want loss and damage to be discussed under adaptation because then they aren't admitting that it's a separate issue?"

The limits to what human societies and natural ecosystems could adapt to were already being broken. Super Typhoon Haiyan was a vivid example of our new reality. In a warmer world, extreme events like tropical cyclones, hurricanes, storm surges, and floods were expected to be more frequent and more severe, fueled by the additional heat energy of the atmosphere and a higher, warmer sea. Other slow-onset events—like the salination occurring in The Gambia, the glacial retreat in the Himalayas, the loss of coral reefs, and the resulting decrease in biodiversity—would also bring permanent losses and damages.

The easiest to visualize was the sea level rise threatening islands like Tuvalu. The president of the Maldives, a chain of tropical atolls in the Indian Ocean, had famously convened his cabinet underwater five years ago. The video of ministers in scuba gear trying to sign bills around a submerged conference table filled my mind. Islands could only adapt so much to sea level rise. Yes, they could invest in restoring mangroves to prevent erosion and shore up ports with seawalls. But they could not adapt to a life that sat under the surface of the water. Nations like Bangladesh and the Maldives would continue to lose territory and, along with it, their cultural heritage and societal identity. And that loss was directly linked to climate change.

Developed countries did not want to talk about the consequences of this. They were liable. The climate change they caused was on track to make whole countries unlivable, to spur death, extinction, and migration on an unprecedented scale. So their negotiating strategy was to not publicly acknowledge that loss and damage was happening. They would negotiate solutions only if couched in the idea that the consequences of climate change could still be adapted to.

"Yes. They can't have it both ways. We can't have low commitments to reduce emissions and no acknowledgment of the damage that will cause," Hafij said.

His iron-clad logic made my mouth pull up at the side in a grin. He was right.

The reductions under the Kyoto Protocol would not be enough, and the ones under the new agreement would not start until 2020. Everyone knew that unless big emitters made drastic changes, emissions would not come down quickly enough to prevent additional permanent losses and damages. We were already seeing them.

"It's sad that even getting loss and damage on the agenda was such a struggle," I told Hafij.

"I know. What we need to decide in Warsaw is how we're going to handle it." He waved over the other LDC negotiators who followed the issue as they trickled into the room. During the preparatory meeting, the team had spent hours huddled together working out what the LDC Group's position and negotiating strategy should be. From what I remembered, members wanted to establish an international organization that was tasked with dealing with the issue.

"Getting developed countries to agree to anything won't be easy," Hafij ended.

The meeting started, leaving me with my own thoughts. I fumed knowing that it was unlikely that developed countries would accept any decision on loss and damage. The liability Hafij was talking about could amount to trillions of dollars in damages—but hadn't developed countries known that and polluted anyway? I tried to contextualize these harms with what I knew of the world, tried to bring it to life. Given that loss and damage was holding the COP's attention, I would likely need to write briefings about it for the new chair and his boss. Tying the concept to my own experiences began with thinking about Hafij's home.

Four years ago, Bangladesh completely overwhelmed me. I arrived on the back of the tropical depression that hit Fiji, the year I traveled on scholarship. According to my guidebook, the Islamic nation's capital city of Dhaka—population fourteen million—had only four hotels that accepted women traveling alone. The rickshaw ride that carried me to one situated in the old town was a journey I would never forget. I could not fathom such a place. There were just so many people. Everywhere. A million colors and smells in the jumble of traffic and script. Words to adequately describe it still failed me.

The hotel's receptionist asked, "May I speak to your husband? He needs to fill this form."

Usually such a statement would have inspired laughter, but he was serious. "I don't have a husband," I said simply.

He craned his head, clearly searching for someone in the lobby. "Your father, then."

It took several minutes for me to explain that I was solely responsible for my stay and that no male relation would present himself. Much of my time in Bangladesh was like this—cultural differences so pronounced that both sides were completely unprepared to deal with the other. For starters, I had never worn so many clothes and felt so utterly naked. Most places I went, I was the tourist attraction, made evident by the legions of blatantly staring men.

I crossed into India in search of tigers, a childhood favorite among the animals. The mangrove jungle of Sundarbans that straddles the border is the residence of the Royal Bengal tiger. I took boats down lazy rivers. Stayed in camps with dancing that started slow and built to a dizzying pace against the sound of drums, flutes, and female voices. Enjoyed bonfires and amazing hospitality. Saw monkeys, what people called a "leopard cat," spotted deer, and wild boar, but the tiger remained elusive.

To ward off malaria during my year of travel, I took the antimalarial medication mefloquine. It had some interesting side effects. The most pronounced, in my experience, were dreams more vivid than any I had ever known. To date, I had been either myself or no one when unconscious. When drugged, I appeared as other people—once a white man, which I found thoroughly unsettling.

Under the mosquito nets of the Sundarbans, I dreamed that a man had the true me cornered. The stares and gropes of the streets pinned me against the side of a building, under a dark sky. My efforts to fight them off were failing. He was going to rape me, and it was going to hurt.

Pain and fear.

"Abba! Strike him!" I screamed in desperation.

Instantly, lightning split the sky and hit the man square in the chest. He fell dead, flat on his face. Smoking, like in a cartoon. The whole scene became so comic so quickly that, still in dream, I started to laugh. I laughed so hard that I woke myself up.

Even so, it was the poverty that I most remembered from my time in Bangladesh and India. The unignorable, unrelenting poverty. In a land of magical food, people were literally starving on the streets. Millions of them. My only references for the diseases I saw were biblical: lepers begged for alms.

My last Friday in the Sundarbans, I hired a rickshaw to take me from camp to the train station. It would be a long ride. On it, I saw a girl of about five standing by the side of the road. She was naked—or maybe she had a string around her waist—and was turned away from me, watching some other children in the palms that lined the pavement. At the sound of our passing, she looked back, her head over her shoulder. She had crap on her bum; it was just *there*. I saw no adults with her. Just a tiny, naked child who had shit herself standing on the side of the road. And she wasn't the only one. And we just drove by.

Everyone I met there lived in the world's largest floodplain. I very much doubted that child had the resources to move. What climate change meant for them was unfathomable, as was the effort it would take to compensate them for it. The crisis would displace millions of people in Bangladesh alone, should its impacts prove irreversible.

The late nights began early with the loss and damage negotiations heating up the COP's first week. Hafij reported all-night vigils to the LDC Group during its coordination meetings, the circles under his eyes growing darker with each passing day. Developed countries were staunchly opposed to establishing anything that might set them up for future liability.

I wished it weren't the case. For moral reasons and for selfish ones too. Working in climate change was stressful enough. Palpable tension in the negotiations pushed things into overdrive, and my wanting to work in the UN didn't make all-nighters any less exhausting.

"We'll keep pushing," Hafij resolved, leaning into the microphone. "AOSIS is standing with us. I encourage all colleagues to join us tonight. We must stay strong together," he finished.

"Thank you, Hafij, for your report," Prakash said from the top table. "I personally will join you tonight. This issue is too important for our countries not to do all that we can."

It was too soon for these kinds of hours! The job of working toward global solutions gave me a sense of purpose unlike any other, but at this rate, I wouldn't make it through week two.

"Malawi is next on my list. Evans, you have the floor."

I switched gears and swiveled to locate Evans, the LDC Group's finance coordinator in the room. He sat behind the Malawi flag. I smiled at Stella, a negotiator who followed the issue of gender and climate change, who sat next to him. I had always admired her brightly colored outfits and had only recently found the guts to tell her so.

"Yes, thank you, Chair," Evans began. "I want to report on finance. Colleagues, as you know, the Green Climate Fund is empty." Evans echoed the Philippines negotiator's words on Monday. "We need pledges in Warsaw. But we've heard nothing. If the developed countries are going to reach $100 billion by 2020, then they should agree to $60 billion by 2016, $70 billion in 2017, and so on."

People nodded around the room.

The plan made sense to me: 2020 was growing closer every year, and negotiators, politicians, and researchers alike were questioning how developed countries planned to scale up their financial commitments from the current flow of roughly $10 billion a year to $100 billion a year in just seven years' time. Midterm commitment numbers felt like a logical way of tracking progress.

"Developed countries need to fulfill their promises," Evans finished, frustration in his voice.

Within the climate negotiations, not delivering on financial pledges was a surefire way of creating distrust, for it demonstrated—in measurable terms—that developed countries were not committed to acting on their promises. Sure, many promises went unfulfilled in international negotiations, but it was easier to claim you had done more or less than you were supposed to when there was no quantifiable indicator.

Money in particular was one that people tended not to forget.

At every negotiation session I attended, the long-term finance discussions were one of the last to finish, and they always featured heated debates. This seemed to be a COP of frustration all around. Marika and I stretched our lunch and dinner breaks out for as long as we could manage, talked about TV shows we liked and celebrities we thought were cute, just to break the tension. I heard shouting around the venue during my trips between negotiating rooms and the LDC office as the hours stretched on.

People were tense.

The *ECO* newsletters, the daily bulletins civil society groups put together that served as newspapers, reported that the Polish government would increase coal production even before building the new plants it had announced, implying that it was purposefully sacrificing progress in the negotiations to suit its short-term economic vision. The protests activists staged outside the stadium grew so loud that I heard them echo within the negotiating rooms. Delegates had to wait until they could hear one another before carrying on.

Temperatures fell as the number of days left until Christmas dwindled. I saw the gray caps and scarfs more and more often. Sandra bought a huge puffy, floor-length winter coat that she only rarely took off. Achala and Sandra spent quiet moments talking about the drain of the negotiating process.

"I'm leaving it," I overheard Sandra say. "I need to find a husband. Have a life. This is my last year."

Achala laughed. "You say that every year. You can't leave. How will I manage without you?"

I began to think that involvement in the negotiations was choosing a lifestyle rather than a profession. And if Achala's dedication was anything to go by, a tremendous amount was necessary to endure the sleep deprivation and long absences from home.

While I enjoyed working with the LDC Group, I hoped that soon The Gambia and the other Least Developed Countries would not be classified as such, that they would have thriving economies with human and financial resources to spare and could fill the halls of the UN with their own wide-eyed researchers. I wished that time would hurry up and get here.

Besides, after two years in the negotiations, the drive that kept me working around the clock was beginning to wane. I wasn't prepared to leave it, though. Not yet anyway, though there remained things I could do without. While overt sexual harassment may have been on the decline, I was constantly baffled by the forthright comments negotiators made about my appearance.

"You've gained weight," was a common greeting I was not fond of. Achala insisted I should take this as a compliment, as in her native Sri Lanka, it was often meant that way. What made it more perplexing was that these comments were normally followed by someone else saying, "Oh, you've lost weight."

I found these speculations maddening and impertinent. My weight had essentially remained unchanged since puberty—apart from that nasty mono bout where for one sickly month I dropped below the 150-pound mark. Anyone who said differently was paying way too much attention to my figure.

"You look like you've been enjoying your food," was a comment Stella liked to give me.

I still didn't know how to respond apart from, "Thank you?"

Prakash surprised me one morning with the greeting, "You have a lovely singing voice."

"I'm sorry?" I answered, completely perplexed.

"You have the room next to mine in the hotel," he said.

Oh no. I was usually an unapologetic shower singer, much to the annoyance of my sister growing up. It had never occurred to me that someday the chair of the LDC Group would know that. I had been singing the *Wizard of Oz* soundtrack for days.

Week one stretched into week two. Ministers and high-level officials began arriving for the political decision-making to come. Among them was Pa Ousman, whom the LDC Group welcomed as their only special climate envoy.

I was happy to welcome him too. His laugh filled the LDC office when he first walked in. Because the COP was being held in a football stadium, the "office" the group had been assigned usually functioned as a luxury box. It had a kitchenette, a minibar, and a balcony that overlooked the field that was now decorated with the white boxes of pop-up meeting rooms. At the spectacle of Bubu lounging in a luxury-box chair to watch the negotiations unfold below, Pa Ousman laughed openly. At his first COP not acting as the LDC Group's chair, Pa Ousman looked more relaxed than I had ever known him.

Once he recovered himself, Ambassador Bhattarai, Prakash, and Achala filled him in on the status of the discussions. Pa Ousman added his voice to the loss and damage negotiations, which were kicked up to ministers for consultation. I was pleased to see such a strong negotiator added to the LDC's ranks. If anyone could help move things forward, Pa Ousman could.

That said, it seemed the LDC Group's hopes for climate finance would go unfulfilled yet again. The only pledge to the Green Climate Fund thus far came from Sweden. Their $45 million was welcome, yet in isolation it was far from the scale of money needed. As the high-level dialogue on climate finance wound down, countries had to settle with Sweden's pledge and policy guidance on the fund's setup as the only takeaways from Warsaw. This, though, was more progress than the discussion on defining the way toward $100 billion. They concluded with nothing except a request for developed countries to think about an approach for a 2020 finance pathway.

After dinner the day before the COP's scheduled end, I reluctantly followed Bubu to find where the negotiations on the 2015 agreement were due to resume.

"Mbaye!" Bubu called, spotting him ahead of us. After he turned around, Mbaye was close enough for Bubu to reach out. Bubu continued in Fula, the language they shared as they walked hand in hand through the building. Senegal was a former French colony and The Gambia a former British colony. Of course, the tribes the enslavers drew borders around had settled in a different fashion. I assumed from the laughter that they were talking about home.

Mbaye waved goodbye when he parted ways. In response, Bubu reached for my hand. I had come to accept that hand-holding was a common thing in West Africa. It must be, given how frequently I saw negotiators from the region walking linked around conference centers.

"Where is the meeting?" Bubu asked as I matched his stride.

"Hmm . . . let's see." I walked toward a live screen. We were standing farther away than Bubu could see, so I read out the room name when the appropriate agenda item scrolled through.

"Thanks," he said.

We walked that way still holding hands. I didn't mind it. It was nice to feel like I somehow tangentially belonged to the Gambian delegation, whose PARTY badge I had worn for two years now. When we eventually reached the room where the 2015 agreement negotiations would continue, Bubu collected the Gambian flag, and we claimed seats—me sitting at the table to hold the chair Pa Ousman would occupy once the discussion started. Ministerial consultations were happening in parallel to the technical negotiations. This translated into lots of waiting.

By 10:00 PM, I couldn't contain my bad attitude. Today was supposed to be our last.

"Bubu," I started whining. "You said COPs end on the Friday of week two. This is my third COP and none of them have ended on time. In Durban, we didn't finish until Sunday. In Doha, it was Saturday." While I spoke, I leaned forward until my forehead rested on the table. "Why are you lying to me?" I pouted for dramatic emphasis.

Bubu laughed at me in response. "You're young," he said. "You're not allowed to complain until you are old like me."

I sighed and closed my eyes. We were stuck on everything related to the 2015 agreement. Nearly every day for the past two weeks, Kishan and Artur had convened us around the square of white tables where my head now lay. It was the largest square of white tables I'd ever seen. Flags were hardly visible from one end to the other.

I had expected negotiators to define things, like what commitments would look like and what sections the new agreement would contain. Instead, I listened to the same debates I'd heard in June, only more heated. Most negotiators emphasized their nation's right to determine the scope of its emissions reductions target. I agreed that ensuring broad participation in the new agreement was important. I just felt that leaving the choice up to individual countries left open too many possibilities for unsafe results.

Sure, the United States could account for economy-wide reductions, but it wasn't setting the bar high enough. Large developing countries weren't counting all sectors of their economies, and some said they didn't have to. Other developing countries said they wanted to work toward doing so, only they needed money, training, and equipment to get there. Some said they shouldn't be expected to do anything without support because they weren't historically responsible for climate change.

Developed countries pushed back, saying that a focus on historic responsibility wouldn't limit warming to 2 degrees Celsius, let alone 1.5 degrees. But when the LDCs and others proposed mandating the scientific community to develop a methodology that kept temperatures within this limit, most changed their tune because they didn't want to be told what to do. Calls for criteria based on equity were similarly thrown out the window.

I was beginning to appreciate how multilayered a challenge arriving at any kind of conclusion was internationally. The consensus-based decision-making process of the UNFCCC meant that a decision was only considered agreed to if no nation objected to it. Decisions needed to be vague enough to accommodate a variety of diverse interpretations and play to a minimum threshold for objection.

In trying to move the discussion forward, Kishan and Artur tabled versions of text that more regularly represented something no country would object to rather than something every country agreed with. This brought the level of the debate and my mood down significantly as the discussions progressed. The texts we were passed were usually brief and overwhelmingly underwhelming. Countries did agree that, in principle, they all shared the goal of combating climate change. That was about it. They all claimed a different portion of responsibility toward achieving that goal. And this left the new agreement too many sides to rectify.

It was such a mess.

The chasm between where we were and how far we needed to go overpowered me. Come Saturday evening, I spotted Hafij in the halls while following Pa Ousman and Bubu to the closing plenary. Kishan and Artur had pushed the conclusion back as far as they could manage, but we were out of time. Hafij told me that the loss and damage discussions had finished in Hail Mary fashion. COP 19 would establish the Warsaw International Mechanism for Loss and Damage associated with climate change impacts. The LDCs and the island nations had pushed everyone to agree to create a forum dedicated to the issue. Though developed countries were against it, they could not deny the facts.

"It's a good result," Hafij said with a small smile when I asked him about it.

"That's great, Hafij. Well done!" His eyes were mostly closed. "Go get some sleep," I said. "You've earned it!" He looked like he was about five minutes away from passing out.

Inside the plenary, nations began to huddle over the 2015 agreement's latest decision text. Though it looked like a replay of the final hours of Durban, I couldn't feel the optimism I had then. Pa Ousman and Prakash reported the phrase "intended nationally determined contributions" back to us when the huddle dispersed. I heard an island negotiator curse as he crossed behind us. The word "commitment" was gone, and he was sure the word "intended" meant we may as well not bother.

I rummaged around my sleep-deprived brain trying to work out the phrase's origin. Commitment was gone because it was too close to what developed countries took under the Kyoto Protocol. We couldn't agree to a division of countries into categories, so we needed a single descriptor that fit everyone, hence, contributions that were "nationally determined." And apparently the United States was only willing to "intend" to do something.

Whatever "intended nationally determined contributions" meant, it was the phrase no one objected to. And the LDC Group, along with the rest, reluctantly agreed to have it rather than nothing. Countries decided to start preparing their intended contributions to reduce emissions under the new agreement as soon as possible. They would communicate these contributions to the international community in 2015.

Now that I understood the issues better, hiding my frustration at constantly having to settle for the least objectionable phrase was extremely difficult. That after all that effort and with so much at stake, the lowest common denominator prevailed. That couldn't be right. As I'd asked in Bonn, if we let countries decide for themselves, how would the UN ensure that enough was done? Would no one be empowered to make sure the new treaty and the national pledges to it were commensurate with the scale of the problem? How was consensus-based decision-making going to effectively combat the crisis we were in?

Bleary-eyed, I watched with Pa Ousman and Bubu from The Gambia's four seats in the plenary as the gavel went down.

10

NEW BEGINNINGS

Kelso, Washington
December 2013

"FOUR FIFTY-TWO, PLEASE. FOUR FIFTY-TWO." I sighed; this was taking longer than expected. Even the automated voice making the announcements sounded bored. I turned around to look through the slowly moving immigration line behind me, searching until I found Noelle, sitting in one of the uncomfortable plastic chairs at the far end of the room. She mouthed, "What number are you?" and I held up four, then five, then four on my fingers, then returned to slowly snaking through the Seattle processing center.

Weeks ago, I had giddily accepted a research position at the International Institute for Environment and Development. For all the pep talks leading up to the Skype interview and the agonizing wait, Noelle was there, offering encouragement. She screamed with me when the job offer came through, ecstatic with my joy. I was thrilled with the career opportunity and the financial security its permanence promised. All the other positions I'd interviewed for after my consultancy had a fixed term, usually counted in months. I loved London, and this would mean moving across the pond. When I finally gained a work visa to the United Kingdom, Noelle and I spent the remainder of that Friday afternoon on Alki Beach—the westernmost point of the city.

It was cold. A deep, face-numbing, record-breaking cold. At negative seven degrees Fahrenheit, it was colder than I could ever remember a Northwest December, likely due to the brightness of the near-cloudless sky. We stood out in it watching ferries on their way to Bainbridge Island and naming the vividly distinguishable mountain ranges surrounding the city. We laughed as we blurred each other's panoramic shots and drank more soy matcha lattes than were good for us.

I counted down the days until my move. I marked them by standing on the cardinal points of the hill I grew up on to take my fill of Mount Saint Helens, white with new snow to the east, the lights in the curve of rivers to the west. My hometown lay before my feet. From this vantage point, the paper mills where the Cowlitz joined the Columbia River dominated the built environment. Trees went into the towers of machines along the water. Paper, paper products, and a kind of sulfated pulp smell came out, fueling our primary industry. A cantilever bridge stretched the width of the Columbia River into Oregon. To my eyes, the land of trees and mountains, water and clouds remained so very beautiful. Home had always been this exact place: the moist air of the Pacific Northwest and the gently sloping, logged hills of Kelso, Washington.

Perhaps it wouldn't always be.

I had dreamed for so long of leaving far behind so much of what this place meant. This place in particular. I sat cross-legged in a slope of waist-high grass, dead for winter. A waving plane of brown blasted into the forested hillside. The rock quarry still had great views. I shivered, wrapping my arms around myself. Even Dad's old hunting jacket couldn't keep out the chill. I looked into the setting sun.

I felt stuck—frozen like the earth.

Hi, Abba, I thought to the wind.

"My love."

Nothing else came and my thoughts were too stretched for sentences. Instead, I watched the lights in town start to come on, glowing yellow against the dark of the valley floor. Above, the clouds illuminated a rose-rimmed sky.

Eventually, I thought, *Abba, do you ever think about Your life?* Then I giggled at how the question would strike someone eternal. *Mine seems a vast, strange journey,* my thoughts continued. *I wonder sometimes whether it will end alone.*

He smiled, *"Will I not be with you?"*

I rolled my eyes. Though a more important consideration, it wasn't what I meant. Behind me, the gaping hole I once heralded as my salvation stretched into the earth. As it had done many times over the intervening decade, the pit reminded me how far we had come and why I should keep moving. Staying here wasn't safe. I should take my latest, greatest opportunity and run into the unknown. Uncertainty overwhelmed me.

"Do not be afraid, my love."

It sounded so easy when He said it, like trust was natural, obvious, easily extended and not easily misplaced. In response, my mind carried me back to that dark night and the months that followed, scraped me down paths I'd tried to seal off and forget or misremember and look away from, determined not to see.

To celebrate my fifteenth birthday, Mom had taken Chanteal and me to dinner at Red Lobster. We sat in a red booth in a Saturday night crowd, nibbling Cheddar Bay Biscuits. My sister wanted to order seafood alfredo, and the waitress wanted to take our order. She hadn't because when we'd sat down, Mom said we were waiting for another person. Mom was waiting. I was not.

Dad wasn't coming.

It didn't matter how many times she invited him, told him, reminded him. Why would he come to celebrate me, the daughter he no longer had? The daughter who didn't exist.

The pain of watching Mom's desperation for another truth was nearly as debilitating as the anguish that had brought me to the cliff edge that January night two months ago.

We sat in that booth for a long time.

Eventually, Chanteal asked loudly enough if we could order that the waitress came over expectantly. After she scurried off to the kitchen, Mom passed me a card in a blue envelope. Below the *Happy Birthday* she had written that her gift was a series of counseling sessions for Dad and me.

I couldn't meet her eyes. Mom had only just come back from her work trip. If she left again, there would be nothing for me here. I didn't want her to be in pain, but she had to know.

"He won't come," I said.

To the counseling. To dinner. He wasn't coming.

She knew that. She must have known that. This pretending had to stop.

Mom started to cry right over the basket of Cheddar Bay Biscuits. She cried so hard that after I gave her my napkin, I saw our waitress turn back to the kitchen. Whatever she was going to ask us could wait. I stared at my mother, drowning.

This was my fault.

Two days later, I stood petrified in the downstairs living room. My parents' house emanated the dark and quiet of a Monday night at 11:00 PM. The only exception was the light and noise coming from the room up these stairs, where my father sat alone watching the evening news, just as he did every weekday night.

Mom was in bed; so was Chanteal. And I had been trying to work up the guts to make the climb for the past fifteen minutes. My heart pounded, and my mouth went dry every time I touched a toe to the first wooden rung.

I was still too scared to go any farther.

My mind labored through a variety of scenarios, each a possible consequence of what I intended to do. I was unsure of which I feared the most. All entailed some form of pain. But then, could things really be worse than they were now?

Enough. If suicide wasn't an option, then there was only one thing left to do, and that was up these stairs.

Don't be a coward, I thought.

Then in resolution, I remembered that I had nothing left to lose.

The upstairs living room of my parents' house was rather grand. Directly across from the top of the stairs was the entrance to the master bedroom, whose door was now firmly shut. To the right, below a half-vaulted ceiling and south-facing windows, was a carpeted loft with views into the kitchen and the downstairs living room. A large television dominated the east wall and emanating out from it were a large sofa, a love seat, two recliners, several end tables, and a series of lamps. Dad sat in his recliner directly in front of the TV. The volume was such that I doubted he heard me coming, so I walked slowly around the large sofa and sat down in his eyeline. I pointedly turned my head from the television to him.

"Dad." He did not look at me. "I want to talk to you." His eyes stayed fixed to the news, as though no one had spoken.

I exhaled. My body was shaking so hard that the breath stuttered on its way out. This had to work. I had to fix this. Dad liked control, so that was what I would give him. I said the prepared words. "I wanted to say, I'm sorry—for everything. You are right. About everything. I want to be your daughter again. I want you to speak to me again."

His eyes shifted in my direction. I stared blankly ahead and swallowed hard against the terror that tightened my throat. *Please, just let me finish.* "I will do anything to make that happen," I promised. Then, I emptied my face of any emotion that might provoke retaliation and apologized for my very existence.

"I'm sorry."

He did not respond. I looked until his eyes went back to the TV. I couldn't think of anything else to say. So I rose. And left.

A letter lay on the kitchen table the following morning. My name penned in Dad's handwriting was on its envelope. I stood looking at it for several minutes before daring to touch it. Eventually, I stuffed it into the depths of my pack and ran out the door to catch the neighbors for a ride. I wasn't brave enough to open it until safely at school.

That afternoon, in the quiet of an empty girls' locker room, I slit the envelope and unfolded the single square of thick, white stationery. It had green trim. I read,

> *Last night I was able to watch you grow and expand into the promise of who/what Brianna will be.*

Repulsion—thick and fast. I looked up and let my eyes blur. It took a significant amount of effort to keep reading.

For when you are able to take the appropriate actions to correct a situation—not just talk about doing the right thing—you mature into a beautiful person. Your mediation skills just improved by taking part in real life. I am proud of you for taking the next step in becoming Brianna.

Welcome back!
Love, Dad

Anger to the point of tears. I wanted to shred the paper and throw it as far from me as possible. What I should feel was relief. It was done. He would have me back. Mission accomplished. Instead, I sat there crying.

Love, Dad.

I had spent my entire childhood wanting that love. His love. I was so desperate for it that I had believed him. I had believed that the chasm between us was my fault, all my fault. But this couldn't be right. Love could not be this hard.

I sobbed until my sides hurt, forcing me to pull air in jaggedly through my teeth. I couldn't do this. I wouldn't make it. It took some time before I could hear the fan above the showers again. Once I could, I counted revolutions. Tried not to feel.

It didn't matter.

"Right" didn't matter. "Fair" didn't matter. Because in the end, my plan was exactly the same. Mom didn't need this, and I couldn't take her pain. I couldn't even handle my own. The hole was dug.

I was fifteen. In three years, I would be eighteen. I could move out of his house, out of his life. Until then, I would do whatever he wished, say whatever he wished, act however he wanted. I would appease him. It was the only way out, the only thing I could think to do.

Grovel or be buried alive.

And then I would leave and never come back.

———————————

The day I moved to London, Dad surprised me by insisting on photographs. He set up the tripod and snapped pictures in front of the drying Christmas tree. Chanteal, Noelle, Mom, he, and I stood surrounded by my mounds of exactly weighted luggage, varying expressions on our faces. Alone in her car, Noelle and I marveled over it. My father did not take pictures of me. During the drive to Sea-Tac International Airport, we talked about this and everything else. Everything, except what was about to happen.

Together we unsteadily tipped my suitcases up the airport's escalators, paid my extra baggage fees, meandered toward the security barrier where we stood looking at each other, not entirely sure what to do. Every other time, there was a schedule for my return, always a known date for reuniting. Now I held a one-way ticket to another country where a permanent job awaited, and I had no idea what that would do to us. The tear-filled eyes of my best friend mirrored my own.

"I'll miss you," I said.

"You always say that," Noelle responded, "but you still go."

"You're coming to visit me," I told-rather-than-asked her for the dozenth time. "You'll like London. It will be great." I tried and failed to work enthusiasm into my voice.

"Sounds good," she said, unbelieving. For Noelle, adventures were real when they were happening. They were real to me from first imagining—I had only to bring them to life.

"It will be. I'll see you soon," I promised.

As they had for the last several weeks, all the goodbyes I planned to say escaped me, their finality too sorrowful to speak. Besides, the words weren't good enough. I couldn't find a way to tell her how much she saved me. Without her friendship, I would have never survived my father's house. I would never have come back. The joy and laughter I found in Kelso belonged so much to her.

"I love you, Noelle." I choked the words into the hug of her shoulder, departure completely and suddenly unthinkable, as was the depth of what I was leaving behind. What the hell did a job matter? How could I leave my best friend?

"I love you too." Noelle gripped my shoulders, holding on. I wished so much that she could come with me, that our laughter could take on a British existence. Instead, our arms fell away and we wiped at our tears. "Bye," was the last thing I said, before turning toward security.

———————

London dawned gray and disorienting. I counted the first week in small victories on my journey to get the United Kingdom to feel like home. On Tuesday I purchased and learned how to use a rail travel pass. I fit in on Thursday by suppressing a laugh when a vendor called me "love." Sunday, I mastered the dizzying number of settings on the washing machine.

I found the city fascinating, the eons of history interspersed with a modern multicultural metropolis, all neatly mapped and labeled. Often the orderliness made a calming counterpoint to my newness. In London clothes were washed one load at a time and hung to dry. Shopping was done in a similar fashion. The pharmacy for toothpaste, the fruit vendor for apples, the baker for bagels. Sainsbury's for everything else. One thing at a time and everything in its place. The ritual of stopping for a cup of tea before introductions certainly made the pace of integrating into IIED delightfully gradual.

I was eager to do well. Not doing so meant leaving the country, which was a level of risk I hadn't fully comprehended until UK immigration marked my sponsored visa with its indelible stamp. Given the amount of effort I was putting in, I wasn't ready to move back across the sea. One of my work inductions was with the pension planner. Thinking about retirement sounded terrifyingly adult. I was twenty-seven. Cup of tea in hand, I sat down in one of the office's smaller meeting rooms across from a man with white hair and a smart shirt.

"You're young," he stated. "You'll be working for a long time yet."

Great.

I watched in shock at the length of the career projections he laid out. According to his analysis, I would still be employed forty-eight years from now. That was nearly twice as long as I had lived. Retirement in 2060. It made my eyes water to think that after another lifetime, I would still be working.

"Let's set you up with some ethical stocks." He clicked through the options on his laptop while my mind turned. "Given that you're a climate change researcher, I'm guessing you'll want a package that's as fossil-fuel free as possible?"

I blinked. It had literally never occurred to me to think about what my money supported. I'd never had any money before. "Ah yes," I said. "That sounds great."

"With the ethical package, some of your largest investments will be in renewable energy. It's higher risk, but as I said you're young—you can afford to be risky. The rest of the portfolio looks to support small and medium-sized companies mainly conducting business in the United Kingdom."

"Awesome," I said, trying to smile. I was suddenly overcome with the urge to investigate what banks invested my money in. Learning about divestment would be something to distract from the loneliness of knowing no one in the country apart from the people in this office.

IIED hired me together with another researcher. Apparently we had both matched the job description so well that Achala decided to grow the team by two rather than one. What no one could have known was how similar we would turn out to be. It would have been hard to pick up on paper.

I was an African American who had lived and worked solely in the United States. Four months my elder, Janna was Bhutanese. She grew up on the French-Swiss border before living in Canada, the United Kingdom, and the United States. She spoke a plethora of languages. I spoke 'merican. She had a lifetime's association with diplomacy and had worked in international development. I had not. Yet upon meeting, Janna and I discovered an uncanny ability to turn up at work wearing essentially the same thing and speak the same sentence at the same time with the same intonation.

It was a bit weird. Imagine meeting the long-lost intellectual twin you didn't know you had.

Janna wore her dark hair just longer than her shoulders and parted it to one side. She favored clothes with back detail and gold jewelry. Like Achala, Janna was slender in the extreme. She held herself with an air of sophistication that was difficult to replicate but could be easily broken, provided you had cute enough cat photos.

We were to divide the writing and project management that accompanied IIED's work to support the LDC Group. This turned out to be easier than I anticipated, given our similar work ethics. We both preferred thinking via typing rather than talking and shared an appreciation for deadlines and clear writing. Dividing the briefings, statements, and talking points and passing them between ourselves became a straightforward, time-saving activity.

Besides, I happily welcomed knowing another person in London. It brought my grand total up to about five. I was elated with the move; it was just proving more difficult than I expected. Janna and I became fast friends, owing both to our similarities and to the sheer amount of time we spent together.

In 2014 the UN convened four climate negotiations. We had traveled to Bonn in March and June. There would be yet another session there in October. And the COP this year would take place in Lima, Peru. Though I knew that countries could call for additional sessions as the need arose, I had never faced the prospect of attending more than two UN climate negotiations per year.

I found the thought of doubling that, well, daunting to say the least. It would mean spending at least eight weeks in the nondescript conference halls of the negotiations, two full months spent working twelve-hour days, six days a week. A bigger cohort at IIED meant more people to tag-team delivery with, and I appreciated working with Janna from our first session together.

The LDC Group welcomed her too at the March preparatory meeting. Thinley and the other members of the Bhutanese delegation knew her family. Some of them talked together in Sharchhokpa, the language they shared. Achala introduced her to the LDC chair, Prakash, and the Nepalese delegation. Then to Pa Ousman, who shook her hand with a smile and thanked her for supporting the LDC Group. Thinking his reception would be the same, I made the connection with Bubu. He surprised me by crying out, "Your name is JANNAH?!"

Janna and I both rocked back a bit at the intensity of his reaction. When she nodded an affirmation, Bubu bowed his head.

"*Jannah tul firdawsi.* A beautiful name. *Jannah* means *heaven* in Arabic," he said reverently. "You are most welcome."

Because our workloads so overlapped, we were often together—so much so that our colleagues inside and outside the negotiations were soon regularly calling us by each other's names, Janna and Brianna seemingly too similar to distinguish. I found this hilarious, as we looked nothing alike. So did Marika. "I think you two should embrace it."

Achala had just asked if "Brianna" could write talking points for Prakash's bilateral meeting, only she was speaking to Janna. Marika had started laughing as Janna looked at me and I looked at Janna and we both looked at Achala, who then realized her mistake.

"You should have a couple's name to save confusion," Marika giggled.

Later, Janna and I branded ourselves "BriJanna" and took to signing joint emails with it. I was glad that working full-time with Marika had expanded rather than limited our friendship. We still found differences of language to laugh over. Back in London, I wrote IIED's communications department with some final corrections to a blog, *Please change the phrase "negotiating block" back to "negotiating bloc." And I missed a period at the end of the second-to-last sentence. Thanks!*

Seated next to me, Marika giggled when the email passed through her inbox. "You missed a period?" she laughed.

I looked at her, completely blank. "Yes?" I had no idea what she was getting at.

"You'll find that punctuation mark is called a full stop," she said sincerely.

"What?!" Now I was the one laughing. "A full stop? That's definitely not what it's called."

"Yes, it is. A period is what you get every month," she protested, which only made me laugh harder.

"That's ridiculous. I obviously didn't forget my menstrual cycle at the end of the second-to-last sentence!" I choked out, both of us in tears. Great. Now comms thought I was pregnant.

Even after working full days in the office together, I still wanted to hang out. Marika opened up her Kentish Town flat to me time and again. Before long, I was regularly inviting myself over for Tom's home-made tacos, which Marika and Tom took pretty well, provided I brought a peanut butter–based dessert and plenty of entertaining anecdotes about an American settling in London. I never ran out of those.

As the tempo of the negotiations picked up, I was increasingly thankful to have Marika's planning and oversight on our side. Efficient and personable people were a rarity in my experience. Marika always had an answer, and her kindness was relentless. She continued to manage the planning and coordination of our work with the LDC Group, handle the finances, and was generally the rock that held our team together.

The camaraderie that grew between Marika, Janna, and me soon changed how I felt about the prospect of doubling the number of negotiating sessions. I wouldn't go so far as to describe them as fun, but the shared experience certainly made them more gratifying, almost like how the trials of tryouts and practice made the sisterhood of team sports worthwhile, win or lose. In the end, you had each other and the work you accomplished together. We even ran the negotiations like a sports play. Marika was on defense in the LDC office, Janna and I ran offense on varying thematic sidelines—she on finance and me on technology transfer, while Achala manned the LDC chair.

I also continued to support Pa Ousman, whose position had changed again. A few months after appointing him as special climate envoy, the Gambian president went on to name Pa Ousman a minister. The ordeal was terrifying. The Gambian president was a former military officer who had held power for twenty years.

Bubu called me in London, frantic. "They have taken Pa Ousman to the State House. We've all come and are waiting for him. Pray," he urged.

Answering the Gambian president's demands or jokes incorrectly was dangerous, to say the least. Pa Ousman was intelligent, but even so, a promotion where a rifle lay on the desk in front of your new boss was beyond a test of logic. Worried, Bubu waited for Pa Ousman all day and gratefully collected him once he emerged with a portfolio that encompassed a dizzying array of issues and responsibilities. His official title was now The Gambia's minister of environment, climate change, water resources, parks, and wildlife.

That was a big deal.

Climate change had never been part of a ministerial remit in The Gambia before. It was something Pa Ousman had insisted on, in a State

House where insisting greatened the threat. The climate crisis was that important to him. Bubu started calling him "Honorable" over email and in person. I followed suit and plummeted through a crash course in diplomacy that kept me lurching from one faux pas to another. Apparently, ambassadors were supposed to be addressed as "Excellencies," while ministers were "Honorable." I still hadn't worked out the formal address for special climate envoys.

I trailed Pa Ousman to an entirely different set of meetings, the level of diplomatic engagement that surrounded the negotiations slowly unfolding before me. His ministerial appointment and past involvement in the UN climate talks gave him the credentials to access any diplomatic forum of discussion. And as much as was possible given his other commitments, he wanted to advocate for the LDC Group. I wanted to help—to support Pa Ousman in the world of climate diplomacy. I signed up for some online graduate classes in international relations, planning to swiftly put them to use. At IIED, I helped secure funding to bring Pa Ousman to any discussion he wished.

As we neared the one-year mark until the scheduled adoption of the 2015 agreement, countries ramped up their climate diplomacy and invited Pa Ousman to high-level discussions all over the world. The Gambian delegation traveled to Berlin for Chancellor Angela Merkel's climate change forum, the Petersberg Climate Dialogue. I thought the UN negotiations were fancy. I had no idea. For the columned halls and table service, I had to invest in several actual suits.

The woman looking at me in the mirror every morning wasn't one I recognized. She paired silk shirts with pencil skirts and fitted blazers, carried an Italian leather computer bag and had a local SIM in her smartphone so people could always reach her. I was called an *attaché*—a word I had to look up. In a chandeliered hall, I sat just behind a round table of ministers, all of whom waited for Chancellor Merkel's appearance so real discussion could begin.

This same setting repeated in Paris and in London. I hailed black cabs in heels and recommended five-star accommodation to diplomats over lunch between meetings at the Foreign Office. I spent months planning a trip to Beijing where Pa Ousman and Bubu met with their counterparts in China. When we arrived, the Chinese ministry literally rolled out the red carpet. They whisked the three of us through a series of meetings and an imperial tour. What they didn't realize, and what I had not yet witnessed in person, was that Pa Ousman spoke Mandarin.

He had been sent away to a Chinese boarding school in his early years and spoke with fluency. Our outing to the Temple of Heaven was marked with Chinese tourists catching sight of the towering, Black trio and running over to take pictures. The women would cry out in joy and shock when Pa Ousman answered in their own language the questions I imagined they didn't think he would understand. The team from the Ministry of Foreign Affairs would shepherd us off once he drew too large a crowd.

The world of diplomacy seemed largely about building and maintaining relationships, which Pa Ousman and Bubu had a knack for. Pairing a refined statesman with a jovial older advisor was a successful tactic. Defining a new climate agreement left them with much to discuss seriously; possessing a readily available bit of comic relief became an asset. The best was Bubu's reaction when confronted by the thousand steps of the Great Wall. He gaped aloud that the Chinese had essentially built an ancient StairMaster. I thought this hilarious. Bubu waited patiently in the parked van while the rest of us climbed between the turrets that evenly framed the green hills.

I brimmed with happiness at my new life. The people I worked with and the places the work took us to, all in the aim of solving the climate crisis, filled me with purpose. I laughed with Abba about it when a meeting took us back to Berlin the evening Germany played in the FIFA World Cup.

I can't believe You sometimes, I thought at Him. Legions of football fans danced in the streets as we made our way to the central hotel, walking distance to the Brandenburg Gate. Fireworks showered the sky. Germany had won.

It's like You're showing off.

The laughter in my mind was deafening. "*Not even close!*"

I go to meetings with presidents. I roll up to cities on the verge of winning the World Cup. People work their entire lives to have schedules like mine, and all of this has just been given to me. Why?

"*Why not?*" He boomed.

11

HOPE SPRINGS

New York, New York
September 2014

I LEANED HEAVILY AGAINST AN ILLUMINATED COUNTER, listening to the receptionist's nails clacking along her keyboard. The automatic doors to my left opened and closed every few seconds, letting in a waft of cigarette smoke from the parking lot. I took in the multiple hues of beige among the uphol-stered lobby furniture, the fluorescent spotlights reflected in the linoleum floor. The lyrics "hotel, motel, holiday inn . . ." started running through my head.

"Brianna and Achala?" the receptionist asked, our passports open in her hand.

I nodded in confirmation.

"There are vending machines and an ATM around the corner to your right. Checkout's at 11:00 AM," she said, handing Achala and me our room keys. It was clear that she meant this in parting.

Achala blinked, unmoving. I watched her eyes sweep the lobby.

"Where's the breakfast room?" she asked.

"You're looking at it. It's those tables to the right." The receptionist popped her gum. "Next," she said waving forward the couple behind us.

I followed Achala's gaze to the dispensers of breakfast cereal lined up along a counter. A swing door in the row of cupboards below was labeled TRASH. She walked over to take a closer look.

"We can't bring a minister here," she whispered. "Is he supposed to eat breakfast with a plastic spoon?"

The side of my mouth pulled up in my amusement at her expression. I tried to see it through her eyes, but I couldn't bring myself to feel the mortification she did.

"It's a Holiday Inn," I shrugged, adding, "in Queens," when this explanation didn't appear to bring clarity. I had known from the description what to expect, and I was too excited to share in her disappointment. Achala just shook her head and turned toward her room.

In mine, I changed into running clothes to shake off the jet lag of the red-eye from London in the afternoon sun. The row houses of Queens were bricked and painted, their back alleys empty and dusty brown in the last of the summer heat. I ran along a cemetery, through a park, and headed toward the sound of the freeway. At an overpass's apex, I wrapped my fingers through the chain-link fence to stare at the tangle of glittering skyscrapers across the water, elated to breathe in the home of my own country, even from atop a six-lane highway. Achala's complaints fell on deaf ears with me. A hotel offering multicolored cereals as a respectable adult breakfast choice filled me with gratitude.

I was back. How I missed America.

———————

It was diplomacy that brought us to New York. September 2014 would host the largest climate-related gathering of world leaders in history. The Major Economies Forum on Energy and Climate—the dialogue President Barack Obama started after the throes of Copenhagen—would headline the summit.

Pa Ousman arrived with invitation in hand.

Since 2009 the group had met to consider how major economies could make progress on climate issues. They began diversifying their list of invitees as the 2015 agreement came into focus. Gathering a broader set of countries meant that any solutions they arrived at would stand a better chance of making it through the UN. That the LDCs were on the invite list was right. No, The Gambia was not a major economy. But a fair solution to the climate crisis required a decision-making process where the poorest and most vulnerable were the standard-bearers of the debate.

In making our New York arrangements, Marika tried her best to book us into a decent hotel. Only, a budget of one hundred dollars a night didn't get you anywhere close to "decent" in Manhattan the week before the UN General Assembly. Not so conveniently for a Least Developed Country, the Major Economies Forum was taking place in the Marriott's East Side Hotel on Lexington Avenue. Even at their special rate for participants, the rooms were triple our budget—hence the plastic spoons of the Holiday Inn in Queens. This seemed to bother Achala more than anyone else, though. Pa Ousman and Bubu weren't ones for ceremony.

My second workday in the city ended with the four of us convened in the fluorescent-lit lobby, going over the schedule I had compiled for Pa Ousman. It was packed. In addition to the Major Economies Forum, he was invited to bilateral meetings, other ministerial dialogues, interviews with journalists, and NGO events.

It would be a busy week for New York generally. To mark the coming year of important international climate action, UN secretary-general Ban Ki-moon had organized a Climate Summit to headline the annual General Assembly. The sixth annual Climate Week was concurrently uniting business, faith, military, agriculture, and health representatives to lay out their visions for low-carbon economies in a series of more than eighty high-profile events planned to take place throughout New York City.

"After the morning session of the Major Economies Forum, I'll go to the Gambian embassy," Pa Ousman said. "Did the Peruvian ambassador confirm our meeting?"

"Yes. It's set for 5:00 PM," I answered. "We're still trying to find a location."

I spent most of my time these days on the phone with aides of other ministers and special climate envoys. Though we came from across the world, all we managed to talk about were which lobbies our respective bosses should meet in.

"I think going to his hotel across from the Marriott will be easiest," I said, making a note. "You'll have talking points for it in the morning."

"Good," Bubu stated. "What's for dinner?"

I laughed.

This question usually signaled the end of our day. Achala tagged along with the Gambians as they headed for a steak restaurant. I packed a bag and got on a subway bound for Harlem.

I'm heading your way, I texted Erina, unable to stop smiling. I drank in the joy of having a social network outside of work, marveled at the prospect of bumping into people I knew—something that never happened in London. Less than a year wasn't enough time to accumulate acquaintances in a city of eight million people. Erina, Michelle, and Noelle had all visited in the intervening months, cheerfully connecting the world I'd left behind with life in my new city. London, love it though I did, simply did not offer nearly enough opportunities for soul food and peanut butter milkshakes.

And there's nothing like your own country: being able to instinctively count out the currency and understand the slang on the radio, knowing where to shop and which side of the road to drive down. It felt great to be back.

I found my way to the brownstone that Erina and her roommates were renting from a professor away on sabbatical, retracing the steps I took during my October visit. The house stood idyllic in its row on 112th Street. I hugged Erina over the threshold. She had just finished her master's in international relations and started a job at the Japanese news outlet NHK.

"So good to see you!" I squealed. "When do Michelle and Andrea get in?"

Michelle was now a physician in training at the Brigham and Women's Hospital in Boston where Andrea was also doing a fellowship. Reunited, we laughed in Erina's high-ceilinged kitchen over the eight months' worth of gossip that had passed since we all shared a room. I fell asleep reveling in the harmony of our matching accents and wishing we could always be together.

Over the weekend, I saw as much of them as I could. Sunday was the only day on Pa Ousman's calendar clear of meetings, and I was pumped to have the freedom to set my own agenda. I spent Saturday night at the brownstone, explaining why we should all be excited.

"We're going to the largest climate march in history," I beamed at Michelle.

"Great. Sounds awesome," she yawned. Her tone was more tired than unenthused.

The three of us were lying on Erina's bed. Andrea was washing up in the bathroom. It was well past midnight. Erina opened her mouth, but no sound came out, probably because we had spent the last several hours in a Chinatown karaoke room.

Word that the UN was negotiating the next climate treaty and that nations would soon come forward with their contributions toward reducing emissions was now mainstream news—even in America. If the 2015 agreement was actually going to result in stopping climate change, politicians needed to make bold emissions reductions pledges, and a good way of getting politicians' attention was to turn out en masse.

Civil society organizations had put the word out months ago, and people were busing in from all over the country. We would convene just before the Climate Summit in what organizers anticipated would be the largest climate change demonstration in history.

I was stoked.

As a Black kid in America who grew up marching, expressing my dread at the state of the world by turning out felt—natural. Just as Granddad had marched and my father had marched, I would march my entire life in the hopes that when my children did, my steps would have brought them closer to freedom. We all needed to mark this under-standing with our presence, time and again. As with racial justice, the struggle for climate justice was far from over. Perhaps dedicated civil disobedience would manifest itself around solving climate change as well. If there was going to be a protest of the climate emergency, I would be there. Michelle's and Erina's enthusiasm for marching or the subject didn't rival my own, but they were good friends, and our changing climate worried them too.

We joined the throng at Seventy-Second Street. Our late night meant that we didn't arrive first thing in the morning. Judging from how packed the holding bays were, though—so full that no one was mov-ing yet—it seemed that showing up earlier wouldn't have made much difference. There were so many people! I checked my phone; estimates had the People's Climate March at three hundred thousand. The crowd stretched for miles in a sea of concerned humanity from all over the country. I was thrilled to witness it, grinning widely while we waited.

I loved the signs.

"Look, Michelle," I said pointing. A doctor wearing his white coat held a placard that read CLIMATE CHANGE IS A HEALTH CRISIS.

Michelle grinned. "Doctors are smart people."

The crowd began to inch forward. We took a few steps, stopped, craned our necks to see above the mass, then took a few more.

When we started moving uninterrupted, we passed the Catholic Climate Movement banner, which read BRINGING "LAUDATO SÍ" TO LIFE IN THE METRO NEW YORK AREA. The parishioners behind it held individual signs shaped as sunflowers. The garden sprouted: CATHOLICS 4 SOLAR, CATHOLICS 4 CLEAN AIR, CATHOLICS 4 BEES, CATHOLICS 4 MERCY, CATHOLICS 4 VEGANS, CATHOLICS 4 BIODIVERSITY. The flowers moved on in slow rhythm.

We marched with teachers, professionals, students old and young, and several polar bears—or at least several people who were dressed as polar bears. Not tied to a group or banner, we moved more quickly than the crowd. Erina had always been a speed walker, and the New York lifestyle seemed to fuel an even faster pace. Michelle and I hustled to keep up. We neared a group of Black and Brown young people in purple T-shirts who were volleying a giant inflatable Earth atop the crowd. Some held signs reading RESPECT INDIGENOUS RIGHTS. The banners CLIMATE CHANGE AFFECTS US THE MOST! and THE NATIONAL BLACK ENVIRONMENTAL JUSTICE NETWORK capped their group.

Ahead, I caught sight of more white coats. Thinking they were doctors, I reached for Michelle then spotted the difference. SCIENCE IS NOT A LIBERAL CONSPIRACY, a placard read. Another written with marker on clear film said, THE SCIENCE IS CLEAR: CLIMATE CHANGE IS REAL. These were scientists, not physicians.

Next to them, a group of parents with young children moved slowly forward—their motivation just so plain. Two moms had the banner PROTECT WHAT YOU LOVE strung between them. A small child napped on one of their backs. One of the kids in the group held up a drawing of the Earth covered with animals and people. Someone had written WE'RE ALL IN THE SAME BOAT on it for him. His father held his hand.

We moved on through the crowd. I spotted a woman with a neon line taped across her T-shirt. Her sign read, NEW YORK CITY SEA LEVEL RISE, 2100, reminding us of Midtown's Venetian future should we continue with business as usual. We neared a group with megaphones holding JOBS, JUSTICE, CLEAN ENERGY signs. T-shirts read, 100% RENEWABLE ENERGY and NO JOBS ON A DEAD PLANET. Signs stated, PROFIT OFF THE SOLUTION, NOT POLLUTION! and CHANGE THE CULTURE, NOT THE CLIMATE.

They chanted, "What do we want?"

"Climate justice!"

"When do we want it?"

"Now!" Erina, Michelle, and I yelled together. We were all so different and so differently concerned. The universality of climate change's consequences hit me again and again. If you cared about anyone or anything on the planet, you had reason to be on the street. No one was untouched.

"I like her sign," Erina nudged me.

"WOMEN FOR CLIMATE JUSTICE?" I asked.

"No, the TO CHANGE EVERYTHING, IT TAKES EVERYONE." She pointed.

The crowd thinned as the collective speed increased. We rounded Central Park and headed south into the city. As we crossed the rise of Fifty-Seventh Street, I looked in hope at the marchers stretching on ahead of us, welling with affirmation. My friends, my people, and my work were coming together. The climate crisis was front-page news, accompanied by photos of Americans turned out for the cause, something I had never seen. We wanted an effective agreement on climate change. We wanted a future we could sustain.

And we wanted them now.

———

In response to our growling stomachs, we peeled off the march at Erina's downtown office. She gave us the tour of her post–graduate degree life, just as I had done for both of them when they each came to see me in London. Afterward we found a French café to sit down for lunch in. It felt surreal that we were here together at this particular time. We ordered and waited.

Erina reminded us that this coming summer was our seventh friend-aversary. She had read somewhere that if you reached this milestone, the relationship would last a lifetime.

"Were you worried?" I grinned. "I think we're in pretty deep at this point."

I touched the silver pendant of my necklace, which had INSIEME, or TOGETHER in Italian engraved on it. We each had necklaces like this with different words associated with our Roman summer.

Michelle laughed out a "How long have you and your boyfriend been together?" On seeing the necklace, a cashier in a grocery store had once asked me this. Rather than explain, I had said, "Three years" and left.

"I still can't believe the part of that story that most upsets you is that I got our anniversary wrong," I told Erina.

She laughed, remembering. "Well, we'd been together four years, not three."

We said our goodbyes as the people's time passed, and the beginning of the week saw the politicians take the stage. For Tuesday's Climate Summit, Ban Ki-moon invited all heads of state to present how they would reduce greenhouse gas emissions and mobilize the political will needed to agree to a meaningful treaty. The emissions reductions pledges leaders announced would form their "intended nationally determined contributions"—the negotiating phrase we had cobbled together in Warsaw.

I was buzzing to attend the UN climate summit. When I ventured into the city on Monday with the Gambians, I realized that there were two things I had not yet seen in New York: the UN Headquarters building standing by the river, the flags of every nation lined before it, and the colossal amount of security it took to make safe a venue where every head of state had been invited.

The lowest on the security clearance roster, I ended up working back in Queens for the Climate Summit itself. I didn't mind in the end. Watching the video feed as over a hundred presidents took the floor and talked through their climate plans had me constantly on the verge of tears. Those who would get the most media coverage were the ones everyone was listening for—Obama, Merkel, the United Kingdom's David Cameron. But after three years of working with the LDC Group, I listened for a different set of voices. The prime minister of Ethiopia pledged to achieve a green, climate-resilient economy with zero net emissions by 2025. The prime minister of Tuvalu announced he would employ 100 percent renewable electricity by 2020. Bhutan reiterated its pledge to remain carbon neutral, maintaining its output of net zero greenhouse gas emissions.

The commitments the LDCs put forward were moving. How was it that those who've already suffered most were coming forward to do more than their fair share? The concluding remarks of the prime minister of Tuvalu stuck in my mind. After announcing Tuvalu's commitment, he stressed that now was the time to be serious and recognize that there was a limit to what the world could sustain. It would be the survival of his people that would prove our collective success or failure.

"For if we save Tuvalu, we will save the world!" he said.

Of the commitments from other countries, those that promised money to the Green Climate Fund caught everyone's attention. Pa Ousman, Bubu, and I had watched Chancellor Merkel pledge $1 billion to the fund in July while attending her climate dialogue in Berlin. Now, I saw France match Germany's pledge of $1 billion, and—in a bold stroke of leadership—Mexico came forward with a pledge of $10 million.

I left New York elated. World leaders were stepping up to the plate, pledging to reduce emissions and delivering on their promises to fund climate action. Perhaps the political leadership necessary to confront the climate crisis did exist. The amount of effort we required was unprecedented. But so was this.

Two months later, President Obama and Chinese president Xi Jinping announced that the two nations accounting for over half of the world's greenhouse gas emissions would act to reduce them. They would also do everything in their power to ensure an agreement was reached in 2015. We heralded this bilateral deal between the two largest emitters as a game changer for the chances of adopting a new treaty. In less than a month, the negotiations would reconvene in Peru for COP 20. France announced its intention to host COP 21 for the all-important final session of 2015.

All signs pointed to success and certainly gave those following the negotiations reasons to hope. Young negotiator though I was, I read them with an unburdened positivity.

Perhaps the UN would deliver the solutions we needed.

Change was coming. I was sure.

12

STALEMATE

Lima, Peru
December 2014

I WAVED IN GREETING as I wheeled my suitcase to a stop behind Marika. She was talking to Janna in one of the check-in lines at London City Airport. Their voices echoed in the emptiness of the early morning terminal.

"Are we ready for this?" I questioned.

Marika smiled and Janna rolled her eyes before saying, "Take it to Peru!" Reaching our first COP working together as a team would require an epic trip. I assumed this caused Janna's mock enthusiasm, rather than the destination itself. The flight from London to Lima was fifteen hours long with an hour layover in Amsterdam. We broke up the journey with mocktail parties by the emergency exit bay, where we planned visiting Machu Picchu post-COP. I was fascinated by the ruined city of the Incas and the opportunity to take selfies with a llama. Achala, who got stuck in traffic and nearly missed the flight out of London, was sound asleep whenever we ventured past her aisle.

Half a day later, we checked in to a quaint hotel in Miraflores. Having gone up a short flight of stairs, I deliberated between my room's two double beds before waking up several hours later. I couldn't remember making the decision.

Morning dawned bright and hazy with the noise of American-sized cars flooding the roadways. A newcomer to Peru, I curiously watched the streets fill before I ventured down to breakfast.

"Look, I'm such a rebel," I said claiming the chair next to Janna.

I had a cup full of coca tea. The guidebooks hyped the brew of small green leaves whose properties were similar enough to cocaine that they were illegal outside South America. The tea bag supplied by the hotel turned out a light-brown cup. I swallowed and waited.

"What does it taste like?" Janna asked, curious.

"It tastes like tea," I answered, disappointed.

"I doubt that," Marika said. She sat down and spread butter on her toast with a disapproving look. I laughed at the face she was making.

"You're so British," I teased. "It's just tea."

After breakfast, we joined Achala in the lobby and set off together to find the conference center. Janna gave instructions in Spanish to a taxi driver who expressed concern that four professionally dressed foreign women wanted to go to Lima's General Army Headquarters. When the taxi dropped us at the base's entrance, he was not the only one confused. People asked bemused questions into walkie-talkies and pointed us in different directions, until a group of soldiers emerged who insisted we follow them to the designated portion of the base. They were clearly entertained by our presence.

I wasn't entirely surprised. A few months ago in Bonn, I had watched the Peruvians' presentation about the COP venue warily. Not only was the venue usually occupied by the army, it also—as demonstrated by the computer-rendered footage of the conference hall—hadn't actually been built yet.

Construction workers outnumbered LDC negotiators three to one. Marika, Janna, Achala, and I climbed stainless-steel stairs into a darkened room, which was difficult to imagine working in, for our meeting. Its canvas ceiling swayed in the breeze and I could hear hammering nearby. Around us the temporary structures that would house the UN climate negotiations were still coming together. As more bodies entered behind us, I realized that air-conditioning units had been installed yet.

With slick hands, I greeted all the usual participants with handshakes, smiles, and cheek kisses. I trailed Janna as she veered off to locate Evans.

When we found him, he shook our hands and asked, "Janna, will you be helping the finance team again this session?"

Janna hedged her answer about the unending time suck that were the finance discussions. "I'll be following gender too, but I'll be with you as much as possible."

Following the financial promises coming out of the New York Climate Summit, the LDC Group was determined to use the momentum to encourage additional pledges.

"Money in the Green Climate Fund will go a long way to helping the 2015 agreement negotiations," Evans said. If anything would inspire the trust necessary to draft the next climate treaty, cash would do it.

I made my way over to where Bubu and Fred Onduri were talking at the front of the room, waving to Hafij as I passed his row. Fred was from Uganda, the country in East Africa that bordered Kenya and Lake Victoria to the west. He acted as the LDC Group's lead negotiator on technology development and transfer. He had small eyes and a fondness for comparing bellies with Bubu to see whose was larger. Bubu always won.

At my greeting, Bubu reached out and took my sweaty hand.

"Are you ready for technology?" Fred asked me. Now that the team was bigger, I could follow these negotiations more closely. Technology was, after all, my area of expertise.

"I am," I said. "I hear that Stella will be joining us too." I scanned the room for the negotiator from Malawi, eager to get to know the pair I would work closely with over the next several weeks.

"Yes, Evans told me," Fred smiled. "The technology team is growing!"

I milled around greeting people, talking to Sandra and Manjeet before wandering over to where Janna was chatting with Thinley.

After the chair opened the meeting, his first order of business was to give the floor to Ian for an update. Tuvalu's prime minister had appointed Ian as ambassador for climate change and environment a few months ago, a move that prompted several long-standing negotiators to clap out cheers of "Ambassador!" while he prepared to speak. Along with his new title, Ian had also taken on the role of the LDC Group's lead negotiator in the discussions on the 2015 agreement.

"Thank you, Chair, for the opportunity to brief the group," Ian began.

He arranged notes on the table in front of him while the cheers died down.

"Successfully delivering a draft text of the 2015 agreement is our biggest priority for Lima." This was the timeline the UN decided in 2012—that parties were to finish the treaty's initial draft this month. "The draft needs to limit warming to the smallest degree possible and hold global average temperatures to an increase of no more than 1.5 degrees Celsius."

I sat dripping in sweat, trying to pay attention to the finer points.

Ian went on to stress the importance of separating loss and damage out from the adaptation section where it was currently housed, as the two issues were both of distinct importance. He also championed the continued recognition of the LDCs' special needs and circumstances as the poorest and most vulnerable nations on Earth.

Though Lima was a major milestone, spirits were running high. New York and the pledges that followed were universally viewed as a tremendous success. The joint announcement by the United States and China was continually cited as a major indicator of hope.

Once the air conditioners were installed during the meeting's second day, I saw the negotiators' ease in the casual pace of their conversations and the quick agreement on LDC positions. While there was much work to do, I left the preparatory meeting buoyant with optimism.

Intent on dinner, I followed Janna and Marika into the early evening sun. While Marika and I kept lists of places to visit wherever we traveled, Janna's preferred mode of exploration was via restaurants. She continued to prove a delightful addition to the team. Lima's streets were warm and laid out in a grid, which made navigating to her recommendation easy.

I had never seen Janna so happy. The restaurant she chose served amazing salads, which came out of the kitchen in shallow bowls overflowing with quinoa and grilled vegetables. "I think she might actually die," I told Marika, feigning concern.

Janna was big into healthy eating. This place had her literally giggling into her meal.

"Poor thing." Marika shook her head. "It's because she doesn't eat dessert. She gets overly excited about the main."

As petite as she was, Janna kept to a strict diet that I found baffling. I shook my head in unison. Across the table from us, Janna finally gave in to our provocation. "At least when I do eat, I order real food," she said, looking pointedly at me. "Who orders soup and a smoothie for dinner?"

"I love a sloshy dinner," I retorted. Janna raised her eyebrows in question.

"You know that sound your stomach makes when you drink a lot of liquid?" I explained, rocking side to side. I felt the pitcher full of *jugo con yogurt* roll comfortingly back and forth. "Slosh, slosh. It's great, isn't it?"

I saw the tide of judgment shift. Perhaps I was more invested in the Nutribullet craze than I realized. I liked the food in Lima: the ceviche, fresh fish seasoned with onion and lime; and the jumbo corn that came as a side dish or that you could buy in Corn Nuts form. I did not like the outrageously sweet drink made from purple corn that everyone insisted we try.

After dinner, Janna, Marika, and I found a church near our hotel that cared for two dozen adorable cats. The church's adjacent park featured dancers that gathered in the warm night air. Some of the women wore white embroidered tops and large pink skirts that twirled when they spun. They completed the look with short top hats over their long, dark hair that hung in thick braids. When they assembled, the dancers joined hands, moving in and out toward the center of their circle to the joyous sound of trumpets. I clambered up onto a park bench the first time we saw them to get a better view, Janna beside me. Eventually we left and turned for the hotel, letting ourselves drift slowly away from the music. I swayed in time, contented with exploring a new city and spending time with people I liked so much.

When we boarded the COP shuttles Monday morning for the start of the formal negotiations, I held every expectation that this would be the best session to date. Everything was at stake. The climate crisis was killing people by the thousands every year. Yet, unlike all my previous COPs, world leaders had just vehemently announced their desire for a new treaty. Now was the moment to deliver on the promise they'd made three years ago to stop climate change.

My mood was slightly killed by the traffic. The twenty-minute cab drive of our first commute had turned into an hour-long shuttle ride. The roads snarled with cars and buses, the number of people on hand for the negotiations seemingly too many to handle. It appeared that the Peruvian way of responding to excessive gridlock was to blow the horn constantly. I laughed as we neared the venue. The government had hired people to dance in Mickey and Minnie Mouse costumes while twirling Silencio Por Favor signs.

Its optimism would not be stifled, and the negotiations started off big. The COP president was minister of environment Manuel Pulgar-Vidal, and his smile and upbeat bounce of salt-and-pepper hair animated the now fully constructed plenary hall. He first introduced a video recording of Peru's president, Ollanta Humala, before a wave of drum-bearing dancers filled the aisles with the liveliest COP opening I had ever witnessed.

I sat mesmerized with the Gambian delegation, which felt different now that Pa Ousman ranked too highly to join us for the technical negotiations. The delegation had grown, though, and included several young negotiators just starting out in government. And of course Bubu remained.

When I craned my head back to find Nepal in the alphabetically arranged plenary, I noticed no chitchat. Someone entirely new to the negotiations had replaced Prakash as chair. He sat stiffly in his suit, and I felt bad for him. Chairing the LDC Group was a difficult job, even for those who knew the negotiations. It would be an intimidating challenge for someone who did not.

The new chair's situation was complicated by the fact that the Nepalese delegation had also sent his boss, a secretary from the Ministry of Science, Technology and Environment. The secretary took over the chair's duties whenever he wished yet vehemently refused the title. It was rather confusing, and it all seemed unnecessarily complicated and bureaucratic to me. I worried that having yet another new chair would limit the group's ability to engage in the negotiations. I certainly didn't envy Achala her task of advising them from her seat behind the Nepalese flag. As a fellow member of the Nepali delegation, Manjeet stuck close to the LDC chair as well. Janna had taken to sitting with Bhutan.

The all-important body responsible for the process of negotiating the 2015 agreement got down to business in the days following. Always together, cochairs Kishan and Artur convened us in a windowless tent in the center of the complex. The cochairs had overseen the negotiations of the 2015 agreement and the intended nationally determined contributions for well over a year now. For every session and at a variety of the diplomatic forums I went to with Pa Ousman, they were there, the faces of the 2015 agreement.

Expert diplomats, they played off each other well. Kishan's apparent friendliness balanced his sometimes-cutting remarks, while Artur's sparing words and precision structured the often-heated debates. Following the Bonn negotiations in June, the cochairs had released a "nonpaper" summarizing the views and proposals voiced thus far on which elements should make up the draft treaty.

"It's merely a drafting tool, a means of getting something down on paper," said Kishan, insisting again that this unofficial document had no legal status.

Its twenty-two pages of bullet point notes under the thematic areas of the negotiations formed the basis for the weeklong special session in October. As they opened the Lima discussions, Kishan and Artur presented an updated version of this paper. It appeared relatively straightforward.

The updated nonpaper took the bullet point notes and formatted them into what I assumed the 2015 agreement would look like. It started with a preamble that laid out shared principles and recalled past decisions. Then there was a section of definitions and another on general context. Then it moved on to thematic sections: mitigation, adaptation and loss and damage, finance, technology development and transfer, and so on, before closing with procedural and institutional provisions.

Most important, these articles were written in full sentences and its pages read like a workable, readable document. Where the bullet points had indicated conflicting positions, Kishan and Artur now wrote them out as options. For example, one paragraph read:

> Option 1: A compliance mechanism or committee is established.
> Option 2: No specific provisions required.

The document was a sort of Choose Your Own Adventure draft climate treaty. So, yes, nations had their work cut out for them. The draft had a lot of options that needed narrowing down. Yet when I walked into the negotiating room, printed copy in hand, I breathed a sigh of relief thinking that at least the parameters were defined.

They were not. While many countries restrained their remarks to pick and choose between the options the cochairs presented, there was no rule precluding nations from articulating entirely new ones. No matter how many calls to logic.

Differentiating who does what under the new agreement was our current sticking point, emphatically argued, especially in the section on finance. Artur sighed as he took over facilitating from Kishan, who had done so over the past several hours. "Sudan is next on the speakers' list, followed by Switzerland. Sudan, you have the floor."

The Sudanese delegate leveled the flag in front of him and pushed the button on his tabletop mic. "Sudan, for the Africa Group," he began, "opposes any text suggesting that all parties mobilize climate finance under the 2015 agreement. The Africa Group recalls the differentiation between developed and developing countries established under the convention. It is the responsibility of developed, not developing countries, to provide finance."

The world had changed since the 1992 convention's bifurcation of developed and developing countries. No one had an acceptable solution for acknowledging that, while at the same time not renegotiating entirely new lists. This made writing the subject of any sentence in the draft text nearly impossible.

"This distinction must be retained in the text. The language 'Developed country parties shall' is the appropriate phrasing."

"Thank you, Sudan," said Artur, hands at his temples. "I would like to remind parties that positioning is not negotiating. I urge you to make precise and concise proposals that can bridge differences rather than restating them. Switzerland, you have the floor."

"As a bridging proposal," Switzerland began, "we suggest including in the 2015 agreement a reference to 'each party' rather than 'all parties.' The language 'Each party should mobilize climate finance' may be an acceptable landing ground."

"Thank you, Switzerland, for that proposal," Artur said. "Delegates, I hope others will build on this in your interventions. Next on my list is Japan."

"Japan would like to see language that encourages 'all parties in a position to do so' to provide finance," the Japanese delegate stated, ending abruptly with no nod to Artur's instruction.

The cochair moved swiftly on. "India, you have the floor."

"Thank you, Chair," the delegate began. "India would like to build on the proposal made by Sudan. 'Developed country parties and those listed in Annex II of the Convention shall' is our preferred option."

"Next on my list is China."

"China would like to support the distinguished delegate from India. China would also like to delete the text suggesting that 'Parties mobilize and provide financial resources in a manner which is capable of adapting dynamically to changing realities and future developments and needs.' This text does not align with the principles of the convention and is incongruous to our efforts here."

Directly across from China, the representative from the EU leaned forward in his chair.

"Thank you, China," Artur said once he had finished. "The European Union please."

"I will speak directly to the previous intervention made by China, as the EU proposed that text," the delegate began. "We would like to clarify that 'evolving responsibilities and capabilities' captures the growth in the levels of prosperity and greenhouse gas emissions of developing countries." He wasn't holding back. "We would also like to note that some developing countries are currently more prosperous than some EU member states," he went on.

Delegates sucked in air. Agreeing on bridging proposals wasn't going well.

Artur eventually took back the floor. "For clarity, delegates," he addressed the room, "we now have five proposals on the table. This increases rather than decreases the number of options. It certainly does not move us closer to agreeing on one."

For hours, Artur and Kishan chaired as delegations proposed new options, some of which varied from the existing ones by only a single word. Every day we heard as many views as possible. Week one slipped by without the cochairs ever concluding the speakers' list. Nevertheless, on the Monday of week two, Kishan and Artur rolled out an updated draft treaty that captured the previous week's discussions. At thirty-three pages, the text had grown to include twice as many options.

Frustrated with our lack of progress, I marched toward the negotiations looking for Ian and the other LDC flags lined up together. I found them at the far side of the room. Surely we were going backward if the goal was to clean the text. We had only four days left to finalize the draft.

"Delegates, let's start with mitigation this morning," Kishan said, kicking off the negotiations. "The European Union is top of the speakers' list carried forward from Saturday. Would you like to begin?"

"Yes, Cochair. The EU believes the mitigation text should reflect that 'all parties will eventually take quantified economy-wide emission reduction targets under the 2015 agreement.'"

Oh boy. Well, we certainly weren't off to a bridge-building start. These were the same positions that I had researched in Bonn two summers ago. If nobody was willing to move, we weren't going to agree to anything. Directly in front of me, Ian turned Tuvalu's flag up indicating that he wished to speak.

"This is of course in line with the goal of limiting warming to two degrees Celsius," the EU finished.

"Tuvalu, I see your flag is up," Kishan said. "Coincidently, you were next on my list."

"Thank you, Cochair. Tuvalu is speaking on behalf of the Least Developed Countries," Ian began. "The LDCs would like to reiterate that the long-term goal must be to keep temperature increase below 1.5, not 2 degrees Celsius. This for us is a matter of survival.

"On mitigation commitments, we suggest the 2015 agreement have two annexes. One annex should be for parties taking quantified economy-wide emission reduction targets as indicated by the EU. The other annex should be for parties that take other forms of commitments, particularly the LDCs, who at this time are not in a position to take on quantified economy-wide emission reduction targets."

When Ian finished, Kishan called, "The United States."

A familiar accent began, "The United States is opposed to creating binary divisions on commitments, based on annexes or the distinction between developed and developing countries."

I hung my head in desperation. This wasn't getting us anywhere. Countries weren't any closer to compromising. From what I had heard over the past week, it felt like they didn't want to.

"Thank you, the US," Kishan was saying. "Bolivia, you have the floor."

"Bolivia, on behalf of the G77 and China, would like to remind the United States and others of the principle of common but differentiated responsibilities as enshrined in the convention. These principles must continue to guide the commitments taken by developed and developing countries under the new agreement. To ignore this would be to not have an agreement."

I closed my eyes.

Twenty minutes later, when Bolivia stopped speaking, a visibly agitated Kishan addressed the room. "Delegates, I feel I must again remind us of the time we have available. If we continue to conduct ourselves in this manner, we will not finish the task at hand."

He was right. I had never heard a facilitator articulate this so clearly. For the first time in my limited, albeit growing, experience at the UN, I realized that many countries in this room did not actually want to agree. After all, in Durban the world had only decided to negotiate a new treaty on climate change. Perhaps that was exactly what they meant. They would *negotiate*. Without narrowing down the options. Without compromise, they would never agree. They would never make progress. They knew this.

What if most governments didn't actually want a new treaty at all?

Days passed with me lost in speculation. Ministers came for the start of the second week. While the fashion on display was enthralling as ever, this year had the bonus of welcoming Pa Ousman as The Gambia's ministerial representative. I was thrilled that his arrival would bolster the LDC's negotiating power.

We needed it.

Marika and Janna looked up from their laptops to smile at me when I entered the LDC office on Tuesday morning. I swept the room for an empty seat, not registering that Janna was laughing. She came over to stand next to me.

"What's up?" I asked, confused by her giggling presence.

Then I registered what she was wearing. It was a solid, mustard yellow dress—just like mine. In an effortless feat of telepathy, Janna and I had unthinkingly coordinated our greeting outfits. She was even eating a banana, just like the one I had in my purse. BriJanna had struck again.

"Nice dress." I laughed.

The office door swung open for Bubu. He said his "good mornings," taking us in when he turned to face our corner.

"Ay! Yellow! What will Honorable think?!" he cried.

Every time I wore yellow, which was often, Bubu reminded me that it was the color of the political party that opposed the Gambian president and made me promise that *when* I came to The Gambia, I would leave it behind.

"You can't wear that to the State House," Bubu went on.

Now that I more fully understood the seriousness of this warning, I planned to comply should I ever visit The Gambia. Surely though, I must be allowed my favorite color outside the country. Bubu raised a hand to shield his face, as if the brightness were too much.

Janna's and my laughter filled the LDC office in response, echoing in Pa Ousman's appearance. Fashion-wise he didn't disappoint. His gold safari suit reached the floor.

"Ay! Honorable," Bubu said in greeting. "Look away," he joked, holding out his arms as though to shield Pa Ousman from the color.

I rolled my eyes. Pa Ousman smiled and shook Janna's and my hands.

"Your talking points for the ministerial roundtable this afternoon," I said, passing him a sheet of paper. The COP president had asked ministers to hash out finance and moving the draft treaty forward. He would convene roundtables on the topics today.

"Thank you."

"Do please make them more adamant," I emphasized. I knew he would; Pa Ousman rarely read my talking points verbatim. "We need to make progress."

Bubu and Janna nodded in agreement.

Pa Ousman went to represent the LDC Group, but the prepared speeches readout in plenary weren't moving us closer to consensus. In response, he met with his peers individually. They shuffled around the conference center exchanging words between hunched shoulders. The 2015 agreement negotiations reconvened by fits and starts in the intervals between the ministerial roundtables and the near-daily stocktaking plenaries called by the increasingly anxious COP president Pulgar-Vidal. Exasperated, Kishan and Artur called for consensus, then rationality, and then just pure restraint. In some cases, a single paragraph ballooned to over ten different options. Things got so bad that parties joined the cochairs in questioning whether, in fact, their peers did want to reach agreement, asking aloud the doubt that had haunted me all week.

Progress wasn't forthcoming in other forums either. Nepal's two-year term as LDC chair was quickly coming to an end. Due to hand over next month, the LDCs needed to decide which would next take on the chairmanship.

As The Gambia had represented Anglophone Africa and Nepal had held the position for Asia and the Pacific, it was now time for a nation from Francophone Africa to take the lead. The problem was that too many countries were interested. More than one had bid, and after nearly a year of discussions, none of the remaining contenders would back down. The seemingly endless cycle of consultation and debate picked up again in Lima. I realized how much I looked forward to the two hours a day when countries weren't fighting. Now, not even the LDC coordination meetings were safe. I found this especially hard.

The LDC Group faced so many challenges that it was difficult to watch them add internal politics to the list. For want of institutional memory, and I think because he liked having her around, the Nepal secretary had Marika join him at the top table. After she put up the agenda, which despite the crucial issues up for negotiation often featured the LDC chairmanship as its primary item, Marika kept a list of those who wanted to speak, passing it back and forth to either the secretary or the chair.

Sitting at the top table during long meetings was the worst; everyone watched as you tried to keep from looking bored. From our seats in the front row, Janna and I did our best to help keep Marika entertained. Twice a day, we made goofy faces at her whenever the infighting went on so long that she started to look sad. When a couple of countries pulled out of the race, the conversation about the merits of the final two candidates went around and around. Though it was easier to tune out, I found not comprehending the conversation more dispiriting when things inevitably switched to French. Unlike Marika and Janna, I couldn't understand it.

Days passed and things were going nowhere.

At the evening coordination meeting, Evans reported news from the ministerial roundtable on finance. Australia and Belgium had pledged millions to the Green Climate Fund, bringing it to its initial resource mobilization target of $10 billion. I looked behind me to find his face, only to see that flags for the chairmanship discussion were already up and Ian, on deck to give an update on the 2015 agreement, had his head in his hands. It was Thursday, the day before the COP's scheduled finish. The climate was in crisis. People were dying by the thousands. Despite years of negotiating, we had an unmanageable draft treaty. The LDC Group couldn't even name its next chair!

Suddenly it was all just too much.

The UN was supposed to stop climate change. It wasn't. The LDC Group was supposed to work together. It wasn't. What was the point of all this arguing? What were we even doing here?

I stood up, walked down the central aisle, and left. I wandered the conference center, opening random meeting room doors until I found one dark and empty, where I curled up on the floor. All the frustration and disheartenment over our lack of progress suddenly broke free.

The LDC Group's internal politics aside, I didn't understand how reaching agreement was possible. I expected Lima to deliver a draft of the treaty: a workable, readable document that gave some assurance that we were headed to a sustainable future, that we had a handle on the emergency we faced. Sure, it could have options, some unresolved questions, provided their answers were framed in the survivable. But options that left open the possibility of warming the world to unlivable temperatures were not options. They were failures.

I had assumed that the desire to agree to a treaty that actually combated climate change was a given. Apparently it wasn't. What was the point then? Why were we trying?

The climate crisis was killing people, and here we sat—for years—arguing over sentence subjects. Negotiators reflected the will of the governments that sent them. And several governments in that room weren't moving. They preferred having no treaty to having one that required them to act. Despite the eloquent words spoken in New York, if political leaders didn't want an agreement, there wouldn't be one.

And the climate crisis raged on. More deadly with each day, emissions went unchecked. I saw faces in my mind, all the negotiators I knew, the billion people they represented, all of whose lives and livelihoods were in imminent danger. The climate emergency was only getting worse. I was too angry to keep from crying.

We weren't making progress; this wasn't progress. Now that a single text was needed, the negotiations of the 2015 agreement were *more* divisive. *More* unproductive. Given that the lowest common denominator prevailed, should countries even manage to agree to something, it almost certainly wouldn't be effective enough to prevent more deaths. The UN wasn't going to stop climate change.

The United Nations was NOT going to stop climate change.

I had left my friends for this, my country. I had framed my entire career believing that the UN was where governments made legally binding decisions that would address the crisis, that the solutions and the international cooperation needed to enact them lay here. If leaders, if governments were not going to solve the climate crisis, what hope was there? The problem was too big, too complex, the threat too great. What should I do?

What *could* I do?

The tears were far from over, but that didn't matter. It was after 8:00 PM on a Thursday evening in Lima, and I was in a Peruvian army base. Why was I even here? On my way out, I didn't bother checking on the negotiations. They didn't matter either.

Friday morning dawned hazy and gray. I debated whether to go in. I didn't believe in what I was doing anymore, had cast my sense of purpose off with the dress I wore yesterday, left it balled up on my hotel room floor. Without clocking the time, I readied myself in slow motion, eventually boarding a shuttle.

In the LDC office, I sat down next to Janna. "Have you seen an updated version of the draft text?" she asked.

I shook my head. "No, not since the one earlier this week."

"I guess this is the final then," she mused. "This is as good as we'll get?"

I shrugged. Apparently, the cochairs had given up too. Our "draft treaty" was now a thirty-seven-page jumble of options. Most gave little assurance that, taken together, they would add up to a climate agreement worth writing.

"There's a new decision to go with it," Janna said. The printer behind her whirled out copies. "Here." She passed me one from the stack.

To accompany the draft, the cochairs had written a decision that laid out what could happen in 2015. The decision included a timeline for finalizing and adopting the treaty. It also spelled out how to submit intended nationally determined contributions and recommended a process for reviewing whether they amounted to the necessary reduction of global emissions to save the climate. The decision was seven pages long.

By noon, the shouting I overheard around the conference center informed me that countries would not accept the cochairs' decision. Even on a proposed way out of this mess, they would not agree. COP president Pulgar-Vidal's smile was gone when he took the stage of yet another stocktaking in plenary.

"My friends," he said in a drawn voice, "we have reached an impasse. To reach agreement, I wish to hear what you want prioritized in this decision. Because we are short on time, my good friends from Norway and Singapore will help me conduct these consultations."

Minister Pulgar-Vidal would meet with individual countries and groups in parallel with the ministers from Norway and Singapore—with luck, working through the list in half the time. "We will hear from you and move forward together," he ended, trying to conclude on a hopeful note.

I made my way out of the adjourning plenary and settled in to wait. When the LDC chair received an invitation for him and representatives from the group to meet the ministers, Marika sent out an email, and Janna and I went out to round people up.

Pa Ousman, Ian, the chair, Achala, Thinley, Bubu, Sandra, Manjeet, Fred, Hafij, Mbaye, Stella, and the LDC Group's other senior negotiators went at the appointed time. When they returned to the LDC office, they reported a productive albeit brief discussion.

The evening stretched on, and Marika, whose coordination work was done for the day, said good night. Janna and I waved her departure. For those of us whose work involved directly supporting the negotiators, our day wouldn't be over until theirs was.

I prepared myself for a long night, with none of the purpose that had accompanied previous COP finales. By 8:00 PM, I couldn't take sitting around anymore and decided to introduce Janna to *The Good Wife* via Netflix, all resolve surrendered. We claimed an empty meeting room and stretched out. In between episodes, we checked in at the LDC office for news updates.

No word on how the consultations were progressing or when the LDC Group would see an updated version of the text. The screens that typically displayed the live feed of meetings just listed CLOSING PLENARY—POSTPONED in flashing letters.

By midnight Janna and I had made it through a good chunk of season 1. When I checked in at the LDC office, I found it empty except for Bubu, who had fallen asleep across a row of chairs and was snoring comfortably.

"Come on. Let's do a lap of the conference center," I suggested after another episode.

Janna nodded and stretched herself upright.

It was 1:00 AM. We headed downstairs from the empty meeting room where we'd been camped out, wandering toward the offices of the COP president. A crowd was starting to form. I spied Pa Ousman, Ian, and Achala huddled among several EU and other delegates.

"What's going on?" I asked Pa Ousman when I got close.

"I've seen China and the United States go in. Others saw India and Brazil. The EU hasn't been consulted yet, and neither have the other groups," Pa Ousman said.

"So no one has seen the new decision?"

He shook his head in response.

I crossed my arms and yawned, prepared to wait with them. By 2:00 AM the congregation had grown. A member of the Peruvian delegation emerged from the COP president's office and told the crowd that the plenary was about to start. People murmured unhappily.

"We haven't seen the text. Why is he calling a plenary?" someone asked.

The Peruvian gave no answer except "Please make your way there."

Never had I seen a late-night plenary so well attended. The LDCs were not the only group in the dark as COP president Pulgar-Vidal took the top table and presented an updated decision, one that reflected none of the LDC Group's priorities.

Discontent rippled through the room as people digested the pages being passed down the aisles. The LDC chair's flag went up in the sea of requests for the floor.

In front of me at The Gambia's table, Pa Ousman shook his head.

"We learned this in Copenhagen. It takes more than consulting the US and China to write a treaty," he said.

The LDC Group would not agree. At 3:30 AM the day after the COP's scheduled conclusion, Minister Pulgar-Vidal adjourned us with nothing.

In the small hours of Saturday, my Peruvian hotel room door closed in defeat. Nearly every bloc had just rejected the COP president's proposed deal. The UN couldn't decide how to get itself out of the mess it was in. And just as nearly every delegate had sounded, I felt listless and despairing. I wanted the negotiations to end. I set no alarm, just drifted off into uninterrupted oblivion. The COP dragging into unprecedented overtime was not new. I just couldn't raise the motivation to see it through. The UN wasn't going to solve climate change.

So I slept.

When I made it back to the conference center, Marika filled me in on the hours I hadn't seen. In a last attempt to advance the stalled negotiations, Minister Pulgar-Vidal was holding yet another round of consultations, with all the groups this time. The LDCs were invited that evening.

"I've emailed the coordinators," Marika said. "Everyone is supposed to meet here in an hour."

The chair convened the remaining lead negotiators and their advisors around a square table in the LDC office. Again, the COP's scheduled end date had come and gone—taking several negotiators home with it. Again, the world could not agree after weeks and months of work. Again, the days had stretched into nights, the weekdays into weekends, and the remaining time ran short. I read the exhaustion and frustration on everyone's faces; it mirrored my own. Some held marked-up copies of the text in hand. Others had laptops and notebooks open, but most had only themselves left to give at this point.

"They cannot leave out the special circumstances of the LDCs," Bubu said. "It's written in the convention and we are still Least Developed Countries. Our contributions cannot be judged the same as others."

"1.5 degrees must be included as well," Pa Ousman added.

"And loss and damage needs to be written in too," Ian said. "We didn't come all that way in Warsaw to give it up now."

I marveled at their ability to keep fighting. I just wanted to go home. They went on until Marika reminded them of the time. She waved them a goodbye as they headed to the COP president's office for the group's appointed consultation, her "Good luck" trailing behind them. Luck.

It was after 11:00 PM when word came of updated text, and I stumbled toward the plenary for what I prayed would be the last time. Minister Pulgar-Vidal took the stage to introduce the document while members of the UNFCCC secretariat handed copies down the aisles. He then suspended the meeting to allow time for blocs to reflect.

I held the new and the old decisions side by side and read through the comparison: holding global average temperature increase to no more than 1.5 degrees Celsius was still listed as an OR next to 2 degrees Celsius; the special circumstances of the LDCs were now included; and loss and damage had found its way in, though in a preambular paragraph referencing the International Mechanism for Loss and Damage adopted in Warsaw, rather than an operational paragraph as the LDC Group had wished. This was weak, but it was better than I thought we would get.

Pa Ousman went off to consult with other blocs and discuss consensus. One o'clock in the morning came and went before we readjourned, too tired, at least for my part, to know whether what we had would be acceptable to all.

COP president Pulgar-Vidal took the microphone. "You have all read the document. You have analyzed it and reviewed it," he began in jittery stutters. "I am sure that you can confirm what I said before, namely that it was built bearing in mind transparency, your opinions, good faith, and constructive capacity. I think this is good, and this moves us forward. I think this is the way that we wanted to act. And I think—and I'm sure," he said, correcting himself, "that we are all willing to rise to the challenge of approving this document. If this is the case and I don't see any objections, the document is approved."

Broken applause sounded, and from the top table, Christiana Figueres, in her role as executive secretary of the UNFCCC, patted Minister Pulgar-Vidal reassuringly on the back. His gavel came down as flags went up. Though they had agreed to it, countries were not happy with the text or the way things had been done.

Ian was one of the first to take the floor. "Loss and damage deserves more than a preambular paragraph." He stressed that the LDC Group felt that the reference to the Warsaw International Mechanism for Loss and Damage was a clear indication that nations would effectively reflect the issue in the 2015 agreement.

Other parties followed to air their own assumptions. I couldn't pay attention to the arguments they were making. Lima would end with every group voicing concern and disappointment.

Sometime after 2:00 AM, we trickled out into the haze, my last walk to the COP shuttles. How different this march felt compared to the stalemate's end in Durban. Then, I had hope in a journey beginning. Now, I struggled to shift the despair of what seemed an unreachable end. The world had an unmanageable draft treaty. Even if, by some miracle, they could finalize and adopt it next year, the new treaty almost certainly wouldn't bring emissions down enough to save us, to end the climate crisis, to protect the vulnerable.

What, then, was the point?

13

LISTLESS

Geneva, Switzerland
February 2015

Given the draft treaty's dire state, the New Year rang with calls for one extraordinary negotiation session after another. In February, the UN reconvened in Geneva. I snapped pictures of the flags outside the neoclassical facade my first morning, face-to-face with history. Based solely on my experience, Switzerland was very cold and not conducive to constructive conversation. You would think that meeting at the venue originally built for the post–World War I League of Nations would inspire cooperation. Surely, they could compromise, narrow options down, make some progress.

Instead, I spent ten days trudging through the chill to listen to speaker after speaker add their own iterations to the text. Even if the options differed by a single word from the phrasing of another, the document had to reflect exactly what country X thought it should say. If every nation carried on like this, there really was no hope. We would need 196 different names for the agreement—let alone the countless number of pages to explain it. The thirty-seven-page draft treaty that came out of Lima ballooned to ninety, with over 550 options.

After a week of bickering, I left the city just as deflated as I had arrived, and I wasn't the only one despairing.

Back in London, I listened to Mom sound exhausted over the phone. When I had said goodbye to my family after the holidays, Dad didn't materialize to see me off. This wasn't unusual. It took Mom saying, "Maybe you should go and say goodbye to Dad" for me to know that anything was wrong.

I found him in the upstairs living room. He'd had just had elective surgery to fix a torn ligament in his shoulder. It was affecting his golf game. When pushed, his posttransplant body broke rather than thrived, and pulls were common. Doctors had told him that there were risks. He wanted the surgery anyway. He looked small—as I had come to understand him now.

His arm in a sling, he stood in sweatpants and told me "Safe travels" in parting.

"I hope you feel better soon," I said.

Since my return to the United Kingdom, Mom reported multiple trips to the hospital. Dad's creatinine levels would spike, his kidneys would shut down, and she would find him wandering incomprehensibly around the house. Each time it happened, he was less lucid and less responsive to her questions, until she couldn't convince him to get in the car. Ambulances were called. He spent weeks in and out of the intensive care unit.

Working full-time and trying to take care of Dad was too much for her now. I felt useless from London, but I struggled to envision what I could do to help. I bet Dad would hate my caring for him just as much as I would.

"He's sitting up and talking. He should move out of the ICU tomorrow."

"That's good," I said.

"Hopefully we'll be back home in time to watch the Seahawks play the Super Bowl. Your sister is coming down. I think we just all need a rest." I wished Mom would get one. But the periods in which Dad was lucid quickly became the minority of the time. The doctors didn't understand what was going wrong.

"Dialysis. A return to the ICU," Mom told me this time. "They're planning to transfer him up to Seattle. They're just not equipped to handle his case here."

"What will you do? Can you take time off?"

"I'll work remotely for a while, stay with Chanteal." My sister was now a librarian outside Federal Way, about forty-five minutes south of Seattle.

"Should I come?" I asked again. I had been asking for weeks. She was exhausted and there was no telling how long this would last.

"I don't want to make you visit," she said, hesitating. I let the line sit quiet until she went on. "I'm afraid if you don't come home now, though, by the time you do, he won't be lucid anymore. He won't know you."

"I'll get a flight then," I said. "I'll call British Airways and see what I can do."

The following day, I didn't cycle to work. And I didn't celebrate what the British called Shrove rather than Fat Tuesday by flipping pancakes with my new flatmates. I flew to the other side of the world, landing that evening in Seattle, where Chanteal picked me up.

"Mom wants you to go to the hospital tonight," she said.

"Tonight?" I asked, surprised. "For the whole night?"

"She's not thinking clearly," my sister answered. "She says she wants us to take shifts. Her and me during the day, you at night."

"That sounds—nuts." Had Mom somehow forgotten how my father and I were together?

"There's no point arguing with her about it now."

"Well, if I'm supposed to stay awake all night, I'm going to need a nap." I suddenly regretted not sleeping on the ten-hour flight.

"OK. We'll wait until the traffic dies down and go to the hospital in a couple of hours. Mom's there now."

Groggy, we slipped through the dark highways of the Emerald City later that evening—exiting at the University of Washington, which for years I had held so fondly as home. We were issued passes with our pictures printed on them and told to stick them to our chests. Chanteal didn't need directions. She knew the way.

I didn't know the writhing man in the hospital bed. He was ninety-seven pounds at last weigh-in, a naked collection of bones. The personality I recognized as Dad wasn't there either.

"Look, Glen," Mom said when Chanteal and I entered. "Brianna is here."

Dad's eyes were focused on an empty middle distance. That wasn't right. My 20/10 vision was a gift from him. I could see as clearly as he could that there was nothing to focus on there. I couldn't tell if he recognized me or not.

"I'm going now. Brianna will stay with you tonight," Mom said, getting up. She hugged me without making eye contact. And then she and my sister were gone.

This was going to be interesting.

"Hi, Dad. How are you feeling?"

My question hung in the air, brokering no response.

I unzipped my puffy jacket, which felt too warm for the Northwest winter, and draped it on a chair across the room. Dad's eyes tracked me as I walked to the window. It faced east toward the University of Washington football stadium that stood clearly illuminated against the night sky. Some time passed in silence. I thought perhaps Dad had gone to sleep. When I turned around to check, he was still staring at me.

"You have a nice view," I said.

He blinked at me.

"Where's Kay?" he asked. Perhaps he did recognize me.

"Mom went home for the night," I answered. "Back to Chanteal's place in Federal Way."

This information registered. Dad tried to shift himself forward, slowly moving his legs as if to climb out of bed. I walked quickly toward him. He looked like he would fall rather than stand should his feet meet the floor. Frustrated by his lack of progress, he eventually held out his hands to me.

"Let's go," Dad demanded.

Understanding what he wanted, I shook my head. "We're going to stay here tonight," I responded. "Try to get some sleep. Mom and Chanteal need a break too."

Then, I saw him.

A familiar anger crossed his face at my disobedience. Dad was in there after all. I watched the fingers of his right hand come together as his arm went up. The movement was too slow to have his desired effect. I caught Dad's fist at my jawline, reminiscing just how long it had been since someone struck me in the face. A lifetime.

We locked eyes.

"Stop," I said.

I held his fist in my hand. A nurse—who I gathered had watched the scene given her timing—entered. I slowly lowered Dad's fist back down to his side.

"Stop," I repeated, anger in my voice now too.

"How are we doing in here?" the nurse asked, businesslike. She moved forward and purposefully tucked the sheet edges back under the bed with a "Glen, let's make you a bit more comfortable." With great efficiency, she gingerly shifted his legs back into parallel lines in the bed's center.

"How about we lower the angle of the bed a bit too. Get you ready to go to sleep."

Dad didn't protest. He seemed unsure of who the nurse was.

"There you go," the nurse said to him once she finished. And then in an aside to me, "There's a nice waiting room two floors up. Good couches. I can sit with him for a while." This was an invitation.

"Thank you," I said, meaning it.

In a darkened room two floors up, I lay down on a couch. It was very comfortable. Should I see that nurse, I would thank her again. Though the sofas of the waiting room's lounge were empty when I entered, I explored every corner before choosing. At the far end, I found a windowless discussion room that closed with a door. Inside, there were just two couches and a wall clock that read 1:17 AM. I guessed that this room heard a lot of bad news. Perfect. I closed the door and flipped off the lights before curling into the fetal position.

God, I hated Dad.

I hated that he was sick. I hated that Mom had left me here to take care of him. I hated that I wanted him to love me. I hated that he didn't. I hated that I was so like him. I hated that he hit me. And I hated that, for so long, hating him had consumed me.

———

The peace that had followed my disastrous fifteenth birthday lasted only six months. Though my response changed, the cycle of conflict picked up again undaunted, even with appeasement as my ultimate objective. The driver's license and car that Mom handed over on my sixteenth birthday helped. Distancing myself became immensely easier.

As did my plan to leave and never come back.

Hating Dad and escaping Dad formed an independent driver of my decision-making, a force in its own right that clenched my jaw and impacted my choices. After I moved to Seattle at eighteen, years went by without enough time with Mom or Noelle. I was so intent on leaving behind my hometown and the man who lived there. And I did love the University of Washington and the friends I made, especially that Divine one, even though the better I got to know Him, the more difficult reconciling the other forces in my life became.

Abba was a God of unrivaled forgiveness. All my reviewing of His story led me to realize that there were no angry excerpts after *forgive those who sin against you* or *love your enemies* that read, "except your father." That made me even angrier. My hate was so clearly justified. What, did He literally expect me to *turn the other cheek*?

Asinine.

But the more preaching I heard and the more chats we had, the more the whole forgiveness thing started to sink in.

I started to see. I carried my hate for Dad around like a brooding cloud of judgment: strangers couldn't be trusted, and friends couldn't be trusted with the truth. Noelle remained the only person I'd told about what my relationship with Dad was actually like. Mistrust cut me off from people, stopped me from really opening up. And if I didn't want that to be part of my adult life, perhaps there was something I needed to do.

I wasn't free of Dad. In the years preceding his transplant, he'd come to Seattle for medical tests, to the only hospital in the state that could handle his case, the hospital at my university. Even given our lack of relationship, Dad had the common courtesy to tell me when he was at my school. Which was how I found myself, several years ago, sitting in a restaurant on University Avenue, about a twenty-five-minute walk from the darkened hospital waiting room I was currently curled up in.

Dad sat across the table from me, and before the bill came, I told him that I forgave him. That I loved him.

It was the hardest thing I had ever done.

He had ruined my life—nearly ended it. Taught me to think myself worthless. And I had hated him and how he made me feel for so long.

I wasn't specific as to what I was forgiving him for, and he did not ask. Nor did doing this mean that I forgot who he was. I knew he was asking for no apology. He never had. He had no plans to change how he treated me. Forgiving him and loving him was to free myself of him. Of the burden of hating him.

Just as I was forgiven.

Just as I was loved.

The mountainous act set me free. Free to live healthy and well, more alive than I'd ever felt. It didn't matter that I needed to repeat the exercise in my mind with shocking regularity. As I now did, the memory heightened with proximity. Again, I called out to Love unfailing. Hand in hand, we would face the mountain of forgiveness. The cost of hope.

————————

The cycle of Dad moving in and out of the hospital and in and out of the ICU continued for months. After my first night, Mom abandoned the idea of us keeping a twenty-four-hour vigil. Instead, we took things in turn. Chanteal came up when she could. Mom and I worked remotely from where we could.

By the end of February, Dad had made great improvement. His creatinine levels were relatively normal, the feeding tube the doctors had painfully inserted was gone. He was conscious, lucid, and doing physical therapy. The doctors wanted to release him. Though they still couldn't identify the cause of the symptoms, physically there was little left that the hospital could do to manage them.

Even so, we knew he wasn't ready. Still under a hundred pounds, he looked corpse-like. Though he would stomach just enough to please a doctor asking questions, he wasn't eating, and he didn't want to be tube-fed. Releasing him would just mean a couple of days' respite from the hospital. We all knew he would be right back in when his kidneys shut down.

And a week later, he was.

His visit started in Kelso's emergency room, where after a quick assessment they loaded him into an ambulance bound for a Seattle ICU. Mom had followed him up north, somehow beating the ambulance to the University of Washington, where she now waited in what would be his room. I remained at my parents' house, set to drive Dad's car up later, though Mom was the only one who believed he would ever need it again.

I stood in the downstairs living room, looking out the window wall at the front yard. Stalling.

"He's going to die, Abba," I said aloud.

"You are all going to die, my love."

Abba's sense of perspective was sometimes annoying. It differed so greatly from my own. "He's dying, now," I clarified with attitude.

"Yes."

Abba, I know this doesn't really apply to You, but do You ever wish You were dead? Death was something I had meditated on quite a lot lately.

I heard laughter in my mind, which irritated me.

Pipe down, Jesus. Not all of us are destined to be crucified for being perfect, I snapped. Then I balked at my lip.

The laughter paused for a second, then restarted at double the volume.

I rolled my eyes and waited until I could hear myself think. *I'm talking a quick, painless exit to the afterlife or to simply walk off with God like what's-his-name in the Old Testament. To never be woken up in the morning by your mother or to have to deal with your family's crap. To be free . . .*

My thoughts drifted off, trying to escape my present situation. They succeeded only for a moment.

I'm so selfish, Abba. Sometimes I wish I would just stop. Forgive me.

"For what?" Abba required specificity when asking for forgiveness. I was not to apologize for being. He did not create me by mistake.

I don't know. Having a crappy attitude. For starters, I'm in no rush to go to the ICU, to see my father full of tubes and hear my mother cry. I want it to be over, I vented. *Surely better than this slow decay. We all just can't do this much longer and yet—what else can we do?*

I wasn't done.

Despite my having just articulated wishing my father a speedy death, all the pent-up frustration of spending months with my immediate family brimmed to my subconscious's surface, Mom's seemingly blind devotion foremost in my mind.

I'm also constantly expecting to be treated well, like I don't know how my family works. They all know. And there I sit expecting it to change. Clearly, I'm the one with the problem. Pain and bitterness soured my tone, defeating me. I hung my head.

Why do I keep coming back?

"Because you love them."

It feels like a curse, a punishment—doomed to spend so much of my life with people who think so little of me. Why did You come back for us? You're so much better than I am. I want to give up. Get on a plane and never return. Forgive me, I thought again before saying, *Amen.*

Let it be so.

In the end, every organ failed my father. Kidneys. Lungs. Heart. He lay in that hospital for another month in a fog of medication. Technically alive but not living. An agonizing decay that was terrible to watch, that I would wish on no one, not even him. I knew death was close when he reached for my hand. And for hours, I stared at the pigment lines running through the nails of his long fingers entwined with their exact replicas in mine.

I had hoped for better between us.

It had never left me through the violence and the pain. All this time. I wanted so much more. For me. For us. A world in which my father and I had an actual, functioning relationship. One in which moments of kindness weren't only possible when there was no time left to give.

He died on Good Friday, April 3, 2015, at 7:33 AM.

Mom held his hand. Chanteal closed his now unseeing eyes. And I sat there, staring at his corpse. Gone. A chaplain came in to pray. My mother and my sister were crying. My tears broke through only when we moved to leave. We would leave Dad in that hospital room.

And we wouldn't come back.

Steady to his wishes to the last, Mom arranged no funeral. I said nothing. Instead, I typed a three-sentence obituary for the local paper and packed my bags for London. In the intervening week, I stared at my journal and wrote nothing. I climbed the roof of my parents' house to stare at the stars and still I could not find the words to say, even to the God I looked for there.

Back in London, I struggled to resume what had been my life. I cycled to work. I attempted to get to know my new flatmates. I attempted to write about climate change. Mostly, though, I thought about how I didn't miss him. We did not have the type of relationship one would miss. In truth, his absence was a relief to me.

But the hope—I knew not how to grieve its loss.

It brought on startling waves of wailing that broke only when I gasped for breath. I spent nights curled into a tight ball desperately trying not to wake anyone, screaming into pillows and coming to in different corners of the room than where I started. I found it the most painful, unexpected emptiness to know that our story ended just as badly as it began. The grief of this realization overwhelmed me. The abyss went on and on—endless in its finality, no earthly way to change its outcome.

With Dad, I was hopeless right until the end.

14

FROM THE GAMBIA
WITH LOVE

Banjul, The Gambia
August 2015

I SAT ON A BEACH looking out at the low waves of the Atlantic Ocean. Desperate to soak up as much sun as possible before returning to the dregs of summer that passed through the British Isles, I had spent most of the afternoon running along the Gambian coast. Afterward, I collapsed to stare at the sea, dripping sweat back at the guardhouse that marked my hotel's perimeter. Clouds so dark they heralded a downpour were steadily approaching the sand. They would soon overshadow the sun. I needed a shower anyway.

The Gambia swelled with a throbbing heat, which I forgave, given the closeness of the sea and the river that cut the nation in two. Bubu and his family had generously welcomed me to their capital several days ago. Insisting I not pay for anything, Bubu carted me around in a two-rowed pickup, blending work and life as we drove between Pa Ousman's office, the market, and the State House, picking up family members along the way.

When I wasn't with them, I typed briefings and ate mangoes while hiding from the monkeys that roamed the hotel compound. I felt completely justified in my dislike. Monkeys were just too humanlike to trust. In my opinion, people should view with suspicion any animal intelligent enough to pickpocket. Their habit of flinging feces wasn't the best either. Before plotting a course to avoid them, I untied my shoes and waded into the shallows until the sea lapped my kneecaps. Wiggling my toes let the gritty sand pass between them in the warm water.

Bubu would likely arrive soon. I was expected for dinner, where his second wife, Kadijatou, would no doubt have prepared a feast. It was nice to glimpse Pa Ousman's and Bubu's lives at home, to see more clearly the people I had worked with for four years now, the people I continued to spend an inordinate amount of time with.

———————

The climate negotiations had resumed in June. We went back to Bonn for the normal three weeks, where the session covered its standing agenda items in addition to the 2015 agreement. The more relaxed pace of business meant there was time to get to know the negotiation's new faces both within the LDC Group and outside it. The LDC chairmanship discussion concluded in an unprecedented compromise that the two-year term would be split into two one-year chairs. These would be divided between the two remaining Francophone African candidates. For 2015 Angola—technically a Lusophone, or Portuguese-speaking, country—would chair before passing off to the Democratic Republic of Congo in 2016.

The LDC chair himself was Giza, a multilingual, charismatic statesman who liked to conduct his meetings over coffee. He came from a family of diplomats and had several years' experience in the UN climate negotiations. His knowledge of the lifestyle translated to his pairing of smart suits with neon tennis shoes. Our team's role supporting the LDC chair would continue, and Janna and I were learning to adapt the talking points we prepared to fit into Giza's free-flowing style. If you wanted him to read something, it was a good move to pass it to him as he stepped out to "talk to his wife," a euphemism that disguised his cigarette habit as matrimony. Relationships were addictive.

The core team of LDC negotiators remained. The preparatory meetings were full of old friends and familiar faces. I heard Bubu's laugh before I even crossed the threshold. I walked in to see Sandra and Manjeet waiting at the front, Fred Onduri talking with Stella about technology development and transfer, Ian and Achala leaning over a laptop. Evans briefed the group on the status of the finance negotiations.

The faces everyone was looking to meet were those of the newly appointed cochairs, the two people set to lead the draft treaty through to completion. Though their decision had ultimately been rejected in Lima, when Kishan and Artur had handed over their posts, the plenary erupted in a standing ovation that was only stopped by Artur's taking the microphone with a deadpan "Please calm down." We had laughed and pulled ourselves together so that work could continue. Whatever anyone felt about the decisions they made, there was no denying the amount of time, energy, and insight the pair had brought to the process. For well over a year, they had embodied the draft treaty. It was strange to walk into the square-tabled drafting room and not see them.

The new cochairs were Daniel of the United States and Ahmed of Algeria. Bonn was the pair's second session on the job and, reading the room, I felt it unlikely they would garner the same trust Kishan and Artur had. That may have had something to do with their national affiliations. The United States, after all, had a consistent track record of negotiating climate change treaties only to abandon them. And Algeria was a member of the Like Minded Developing Countries Group, the negotiation's newest bloc, formed to uphold the convention's original principle of common but differentiated responsibilities as well as developed countries' historical responsibility for climate change.

Or it could have been their diplomatic styles. Dan was well known in the process, an American familiar with the jargon of the talks. He spoke well yet never managed to say much of substance, which, given his background, could not be trusted. Ahmed was new to the UN climate negotiations. He looked down over his glasses to deliver one-sentence instructions from the top table that were utterly impossible to understand. Declarations like, "The text being what it is, we proceed as follows," were given with no further explanation. The things Ahmed said in response to pleas for clarity just raised more questions. He also had the annoying habit of taking selfies with his smartphone every few minutes.

Dan and Ahmed had taken several opportunities to publicly lower the bar for the expectations of what the treaty would do, a dispiriting sign from the people overseeing its negotiation. In an August interview that aired just before I boarded my flight to The Gambia, they said that it was *not* the 2015 agreement's ability to limit warming that would mark its success or failure.

What?! What were the UN climate negotiations for, if not to ensure that greenhouse gas emissions came down? It made no sense to me.

The cochairs argued that a successful agreement was one that established a process for all governments to increase action over time. So rather than solve the crisis now, the UN would set up a process to do something later. Real useful. I read this as a diplomatic cop-out for the sake of claiming "success." Surely, we hadn't put in all this *time* and all this *effort* to achieve action *someday*. The LDCs and everyone else dying of climate change needed action *now*. I sighed thinking about it, my toes still submerged in the Atlantic. When the rain began to fall heavy and fast, I sprinted off through the hotel compound.

———————

The Gambia was hosting the Cartagena Dialogue, a long-standing discussion group that brought together similarly interested countries outside the negotiations—giving them space to find commonalities between their respective blocs. Arriving early had meant extra time to meet and see the domestic side of Pa Ousman's climate diplomacy work.

So, apart from dinner, Bubu was collecting me for another interview. I wanted to know about The Gambia's national conversation about climate change: what the media covered, what information the government was collecting, what people actually talked about. Bubu was happy to further my understanding. Anything that gave me context for Honorable's speeches and had us exploring Banjul was an outing he cheerfully arranged.

When we arrived at Gambia Radio and Television Services later, Bubu introduced me to the baseball cap–wearing chief executive who showed me around the station and answered my plethora of questions.

"Every village collects climate information," the executive said, pausing for breath between my rapid-fire queries. "They track when the rains come and where they can find fresh water, how the trends are changing. The information is broadcast in local languages."

I nodded, wishing my own country would regularly air such vital detail.

"We also film climate change events like the COP and interview Honorable Pa Ousman before and after," he continued. We went on to talk through how these interviews could complement Pa Ousman's diplomatic strategy, only leaving the station in the early evening. Security guards raised the gate so Bubu's driver could exit onto the dirt track. We hadn't gone far when Bubu instructed him to pull over.

On the roadside, a woman sat rotating foil-wrapped lumps in the coals of a small fire.

"How many corns should we buy?" Bubu asked as he and his driver opened their doors. I leaned out of the truck's back seat window to watch them converse in what I assumed was Fula. The road was quiet; the wind moved lazily through the trees on either side.

"Try one," Bubu said, passing me an ear. I hadn't taken more than a few bites before his "Have another" followed. I waved it off with thanks, trying to remember if I'd eaten breakfast that morning.

Bubu turned to stare in concentration at his hands, audibly counting out the number of people currently at his house. He quickly ran out of fingers, gave up, and bought the woman out completely. Bubu was, by Gambian standards, a wealthy man. The vendor grinned wide, handing him two full plastic shopping bags.

When Achala landed a few days before the Cartagena Dialogue began, Kadijatou surprised us both by visiting our hotel with a slender man in tow. He was a tailor and would be making us a gift. "You both must have a dress," Kadijatou exclaimed. "Here, look at this catalog."

Achala and I bent over the glossy pages she thumbed through, pointing out the different styles. "There are so many," I marveled.

A variety of shapes and patterns went past. Whatever was chosen, I knew I would not look as fabulous as Kadijatou. Like the other Gambian women I knew, she wore a long, fitted skirt that came in at the knees before flaring out in a mermaid cut. This was paired with a tailored top, large earrings, and a matching headwrap. Her outfits were of bold colors—pinks, reds, blues, and oranges—all emblazoned with patterns and embroidery.

"I think this one for you," she told me.

The tailor moved forward unfurling a tape, the gesture triggering the realization that these clothes were made to measure. Starting at my knees, he charted six circles around my body. So not only would this ensemble be custom but it would also be snug. Achala laughed at the expression on my face as this knowledge took hold.

Participants of the Cartagena Dialogue arrived in groups, slowly populating the hotel compound with representatives of the world. The UN was still calling for extraordinary sessions to narrow the treaty's exploded draft into a workable document. They would reconvene in Bonn next week. Before then, Pa Ousman wanted to rally progressive allies, find a way to move things forward. I remained despairing about the negotiation's prospects, yet helping the Gambians was second nature now. I ran around with briefings and talking points, tweeted pictures of the growing convening of diplomats.

The Latin Americans, so boisterous in their opinions and outwardly friendly, had delegations headed by young, fierce-talking women. They made such a welcome change from the suited men that dominated most delegations. The Europeans and the Nordics arrived together. They included a Swedish ambassador who spoke Spanish with the Latin Americans late into the night yet was up at dawn swimming in the hotel pool with her blond hair tied up out of the chlorine. The Caribbean delegates included a pair of well-spoken, dreadlocked women, one of whom towered over the rest. The French were eager to build bridges now that the Paris COP was only a few months away.

The dialogue's first morning, I watched Pa Ousman welcome the twenty or so ambassadors to The Gambia. The principal items on the agenda were to identify convergence areas in the draft treaty and discuss how to raise its ambition. They talked about the reasons for hope.

"The United States and China have submitted their contributions well ahead of the deadline, and the EU came forward months ago. It's the best signal we could have got."

"But what's in them! Combined, the contributions won't amount to the emissions reductions we need. We must write into the agreement that successive commitments have to go beyond the previous, and we need a compliance mechanism."

"Are you kidding? You know the US is barely willing to report on their efforts. Getting the principle of no backsliding into the agreement may be possible, but a compliance mechanism?! Please."

Hearing them talk so freely made a nice change from the layered diplomat-speak of the negotiations, even if what they said meant we were doomed. Countries were supposed to share their intended nationally determined contributions as early in 2015 as possible. It was August, and twenty-five nations—including the European Union, the United States, and China—had. What analysis now showed, however, was that the world's three largest emitters weren't promising enough action to hold global average temperature rise to below 1.5 degrees Celsius. This was no doubt what had inspired the cochairs' comments that both lowered expectations for the 2015 agreement and postponed the work.

If the agreement was to succeed in limiting warming, we would have to build in some sort of mechanism for increasing emissions reductions over time. Perhaps to demonstrate the urgency of this need, Pa Ousman ended the day by walking the group of delegates down to a particular patch of beach. I had puzzled over its uniqueness from the countless other beaches I knew all week.

Just beyond the hotel's perimeter fence the sand gave way in a sudden drop. It was a good five feet down to the sand at sea level. You either had to jump for it or find a wooden ladder to descend. At the bottom, white tarps of giant sandbags stacked one on top of the other shored up the level change. Still separating you from the waves was a concrete wall some twenty feet away. This you also had to scale to reach the ocean. I assumed that when the hotel constructed the wall, it met the elevated ground.

Pa Ousman was showing the ambassadors what climate change was doing to the Gambian shoreline. They were losing territory to the waves. Fast.

"The sand has been washed away," Pa Ousman said. "We've done several projects to reinforce this beach, but they've been ineffective. This sand," he said, stooping to let a gritty handful run through his fingers, "was imported at tremendous cost, and we're still losing the coastline."

He paused, looking at the line of other hotel compounds stretching on from ours down the shore. Most had some form of coastal protection, and most were nearly empty. "The beach is the backbone of our tourist industry. And it's failing."

He continued to explain the coastal erosion projects The Gambia had invested in. With sea level rise and changing weather patterns, most of the newly imported sand was being carried back out to sea.

A Dutch ambassador started asking technical questions. "The Netherlands has done several of these projects to save our coastline; which shoring systems have you tried?"

I stopped paying attention and instead looked at Pa Ousman and Bubu standing on the beach. I knew that the livelihoods of The Gambia's population and the bulk of its economy were tied to rain-fed agriculture, that millions of their people depended on it to survive. I knew that the delta country surrounded the Gambia River and that sea level rise's increase of groundwater salinity would affect the entire nation, that two-thirds of this city was less than a foot and a half above sea level. The climate crisis would hit their way of life on two inescapable fronts. The rain would come less predictably, and the water would rise.

And Pa Ousman, Bubu, and their families would have nowhere to go.

The next day, Kadijatou presented me with a two-piece outfit that featured a sweetheart neckline, ruffled sleeves, and an ample flare at the bottom. She spread it across her bed for me to appreciate. The fabric was green with quarter-sized circles of red and blue with yellow and white stripes running through them.

"Thank you so much," I smiled. "It's amazing."

"You are a real Gambian now," she said in response.

That night we sat around her living room eating mangoes and roasted corn that Damba, her eight-year-old son, handed out one after the other. I fanned myself intermittently. Bubu had no air conditioning despite the year-round constant of ninety degrees Fahrenheit and 80 percent humidity. A slow train of ants made steady progress through the sparsely furnished dining room, no doubt in search of mangoes too. Afterward, Bubu's driver took me back through unlit streets where single men walked through moonlight, passed homes without lights or built floors.

I was not a real Gambian. Kadijatou's welcome was genuine and being an African American in Africa was certainly fascinating—I kept spotting familiarities between the face and body shapes of my relatives and those of people living a continent away. I didn't need reminding that Africa was once home, though more blatant reminders were available. While on a boat in the Gambia River, Bubu had asked if I wanted to visit Roots Island, to see for myself the British torture chambers turned memorials to enslaved people. I couldn't face it. I was terrible at genocide tourism. I knew from experience that I could stomach museums but that the actual sites were too much. I couldn't set foot in Hiroshima or the killing fields in Cambodia. People hacked to death with shovels. Children in cages. I couldn't board the bus to Auschwitz. The slave ports of West Africa would destroy me.

"What's the problem?" Bubu asked. "Will you cry?"

"No. I'll vomit," I said with perfect certainty.

What I had only imagined for so long, suddenly real, tactile. Smellable. The millions chained along lazy rivers, bound over the sea to die of forced labor in America. My family. Did my relatives hail from the Gambia River? We had no idea. Enslavement left no trace, no history, any direct link with the continent taken from us.

Then there was climate change. I was certainly not Gambian. I was not from a Least Developed Country, and their struggles were not my own. My only experience of poverty was the lean AmeriCorps year after undergrad, when the end of the month meant a SNAP refill. But food stamps meant a choice of *what* I could eat, not *if* I could eat. My access was so different from theirs. And I had emitted more carbon by the time I was an adult than Bubu's innumerable family would emit in their entire lives.

I didn't have to be Gambian to see this for the injustice that it was and know that there was too much at stake to let wealthy countries waffle over the 2015 agreement. I felt devastated that ambition was now understood only in the future rather than present tense and that even should the UN manage a treaty, it wouldn't be enough.

I came back to the beach the night the Cartagena Dialogue concluded, and one by one its participants flew off toward Bonn, where another negotiating session was to begin the next week. This wasn't even the last scheduled trip to Germany before December's COP. Yet another round of negotiations was planned for October.

After dark, I wandered out to sit on the beach again. Lord knew I wouldn't be sleeping anyway. I navigated the various level changes between the hotel and the sea before plopping down on an empty patch of gritty sand, arms around my knees.

Curled small, I stared into the sea, trying to feel nothing, see nothing but the waves. I used to have so many words. Now, I so often felt like a specter of my former self. I didn't call. I didn't write, didn't keep up with people. I cried myself to sleep for reasons I chose not to articulate, only to dream of death. Nightmares ended with me on the floor, after which I couldn't face breakfast.

I wanted to forget about the negotiations. Their inaction stood in deadly contrast to the inundated capital where I now sat. They had become a kind of endless drone, now almost meaningless for their duration. Like the TV shows watched so often, they were really just background noise. I didn't want to think about the ridiculous amount of time, money, and carbon emissions they devoured. And for what? Given my current proximity to the acute need, sitting on an eroding beach, I couldn't see how what we did at the UN mattered. The point of it all escaped me.

I closed my eyes and listened to the moon pull at the ocean, lost again in the abyss. I was so ready to close this chapter of my life. My father was dead four months now, gone and not coming back. The portion of our relationship that existed died with him. He was over. It was over. Then why couldn't I do this? How long would this listlessness go on? I wanted to feel like myself again, to find new reasons to hope. There were answers to both impossible situations. Letting go of what was now gone would probably take professional help, but I needed to try. I needed to break the silence and open up, reach out to my friends, be a good friend, help and be helped. To stop facing the crises of life alone.

I thought about my colleagues and the work we did together. Quirks, foibles, and all, Pa Ousman and Bubu were two of the best men I knew. They had told me what they hoped The Gambia would hold for their families' futures. They had struggled so hard to make change, worked for so long. I thought about Thinley, Sandra, Manjeet, Hafij, Mbaye, Ian, Stella, and the other LDC negotiators, all of whom had people of their own who inspired such an unheralded effort day after day, year after year. How different their worlds were from mine yet how similar the things we valued.

The UN was our best hope of getting governments to stop the climate crisis. If a new treaty could deliver some piece of that, wasn't it worth seeing through?

15

THE TRIUMPH OF PARIS

Paris, France
November 2015

I SAT IDLY CHATTING with Janna in the Eurostar lobby, the international departures portion of London's St. Pancras train station. We were both slumped over our luggage in the early morning, half-focused on the screen that would show us the correct platform when the moment came. Rather than the long-haul flight typical of COPs, we would accomplish this year's venture with the two-hour train from London to Paris.

I was exhausted, and it was more than just the early start. Though I technically lived in London, I had spent more time outside the United Kingdom than in it this year. Five negotiating sessions, several diplomatic missions, and months in Kelso had depleted my capacity for exertion. This made stepping onto the train platform feel strangely akin to walking into a final exam or that dreadful morning of waiting before my master's defense. I had nothing left to give, but I still had to face the panel—only, my student self was much better prepared for what faced it than the world. And we weren't off to Bonn for another week of trying. The next three weeks held the deadline that the last four years of negotiating had built toward.

"Can you believe this is it?"

"Not really," Janna answered. "It doesn't feel real yet." Her words hung in the air as the train pulled away. Given Paris's proximity to London, the IIED team would come and go in waves rather than en masse. Marika would meet us at the end of the preparatory week. She had booked a hotel in the nineteenth district along Canal Saint-Martin and sent us off with well wishes. Janna and I would meet Achala there, along with most of the lead LDC negotiators who—given the popularity of this session—had gone to Marika for help booking a place to stay. When we arrived at Gare du Nord, I followed Janna through the bustle of wheeled luggage and French commuters to the taxi stand. A few words from her sent the driver east into the city. Janna once told me that charming the French was easy. They never expected her fluency, so she dazzled from the moment she opened her mouth.

The taxi drove us past the iconic grand facades and cast-iron balconies, the cafés and stylish population who dressed in black no matter the season, and finally, down a side street by the canal whose sidewalks ran right up to the line of water. The driver stopped in front of a set of glass doors.

"*C'est là,*" he said. "*Bonne visite.*"

"*Merci beaucoup,*" Janna smiled.

When we stepped out into the November cold, I shivered. With the UN's deadline for delivering the treaty looming, I expected the upcoming negotiations to be intense. Three long weeks—and likely several overtime days—stood between me and a well-deserved Christmas break. I prayed to make it. All I wanted was to go home.

I trooped downstairs for breakfast the next morning to find the LDC chair, Giza, talking with Achala and Janna over baguettes and butter. As I made a plate for myself, the faces I had seen more than any others this year started congregating around their table. Ian greeted Hafij, who greeted Fred Onduri, who greeted Mbaye like the old friends they were. Over the course of 2015, we had spent more time together than apart.

The LDC Group had never tackled a COP staying in such close quarters, but Giza seemed happy to have his core team close at hand. We had all taken tables at the far end of the dining area and the LDC chair pitched questions around as they came to him.

"Ian, what about the ADP coordination?" I overheard Giza ask in the morning chatter. Ian was overseeing the group's strategy for engaging with the body responsible for delivering the 2015 agreement. "How shall we arrange things?"

Giza then pushed away his empty plate and reached inside his suit pocket for cigarettes. Ian was halfway through answering when Giza cut him off. "Good. Good. I suppose we'd best be going—I'll just run out to talk to my wife."

It almost made me laugh. Giza had a way of letting you know immediately what he thought of your answer. If he knew that you had a plan and felt confident about it, he would much rather smoke than hear the details.

The LDC Group's preparatory meeting was taking place at the UNESCO headquarters just south of the Eiffel Tower. We set off on the metro across town once the chair finished smoking. I loved systems of public transportation, even in languages new to me. They were so wonderfully democratic. Anyone could go anywhere. You just had to connect the dots along the route, follow the signs, and find your way. I was aware that this love wasn't shared by all. Achala and Giza bumbled along behind the group, missing us entirely when we changed trains. We stood waiting at the correct stop until they caught up to us, exiting the metro later than planned. Together, we walked toward the UN flags flying down the street.

Even gauging from the number of people waiting for us at the LDC Group's preparatory meeting, it was clear that this COP would be different.

Not even 10:00 AM and the room was at capacity. Well over two hundred faces filled the place, most of whom I'd not seen before. The LDC Group I had come to recognize over the last four years had tripled in number. I looked to find the faces I did know—waving when I spotted Bubu, Sandra, Stella, and the others as I followed Giza, Janna, and Achala through the crowd. Giza adopted a statesman-like persona, shaking hands and greeting people before he took the top table with the other members of the Angolan delegation.

He switched on a microphone and called the meeting to order. Last week, I had puzzled over writing his opening remarks, drumming my fingers against my desk, struggling for phrases to introduce such a moment without betraying my disbelief. I exhaled as Giza began, delivering the words I had settled on. He spoke of the tremendous importance of this moment in history, how Paris was the culmination of the work begun in Durban.

"Without the leadership of the LDC Group in those final tense moments in 2011, the process to negotiate the 2015 agreement would have never begun."

This was what I chose to highlight, the knowledge that kept me here. It didn't matter whether I believed the UN's efforts would result in a treaty capable of stopping climate change. I was here to help the LDC Group, which represented the most vulnerable and least responsible, to negotiate well—as it had for the past four years.

I relaxed when he paused for applause.

The LDC Group had reason to celebrate. Despite their minuscule emissions, forty-two of the forty-eight Least Developed Countries had submitted intended nationally determined contributions ahead of Paris detailing how they would act to reduce them. As Giza said, "If the LDCs can act with ambition, so must ALL."

COP 21 would be the true test of this. The world would watch with expectation; every nation was meant to sign off on a new agreement and sign up to a future of emissions reductions.

"It is now time for the LDC Group, as one of the key proponents of the 2015 agreement, to see this process through to a successful conclusion—the adoption of an ambitious treaty," Giza ended.

Applause filled the room again.

I smiled at a job done well, another speech, another rallying cry. Now for the technical work. The draft treaty that came out of the October session was almost half the length of the ninety-page monster we had grappled with in Geneva. Coming into Paris, we had a fifty-four-page draft agreement that was organized into eleven articles. These covered the main themes of the negotiations and were broken down into a logical structure. The draft I carried around had a cheat sheet written in the margins: *Article 3, mitigation. Article 4, adaptation. Article 5, loss and damage . . .*

For the past year, the negotiations had divided into spin-off groups, each responsible for one of the eleven articles. It was impossible for any single LDC delegation to follow them all, so coordination between the different countries that made up the group had been key all year-round. Finished with his opening remarks, Giza explained this to the room filled with new faces and divided us up into discussion groups covering the different elements.

"Fred Onduri of Uganda and Stella of Malawi," he called, "put up your flags."

Giza went on as heads turned in their direction. "Fred and Stella will lead the discussion on technology development and transfer. If you would like to follow this issue, please join them."

After Giza finished the list of eleven, I moved toward the technology circle, where my expertise and interest lay. About twenty negotiators gathered in the corner where Fred stood waving, turning and dragging chairs into a circle. The longest-standing negotiator, Fred opened with a short welcome. It ended with, "Brianna, take us through the text."

At this point in the year, this expectation no longer caught me off guard. I had lost count of the number of times I had done this. It was why I was still here. I wanted the LDC technology negotiators to be well briefed, and I could help.

"All right," I said clearing my throat. "Hello, everyone. Technology development and transfer is Article 7 in the draft text and starts on page 18." I paused for those with a copy to flip or scroll to the correct place. "We have seven paragraphs, beginning with a goal for environmentally sound technology development . . ." I continued. Members of the circle leaned forward and began making notes as they followed along. Stella and Fred added to my summary as we brought those new to the negotiations up to speed. We told the story of how these paragraphs had come to be, starting with the country that had introduced the idea and explaining how the proposals had changed over time. Stella asked those in the circle to share their thoughts. Their questions surprised me.

Most of the LDC delegates who followed technology were scientists or engineers. Stella was a forester. When they were not at the UN climate negotiations, their jobs were fairly technical. Their governments had sent them to negotiate for better weather monitoring systems, drought-resistant crops, or ways to increase their hydro capacity.

Today's questions were about process and protocol.

"Who represents the G77 in the technology negotiations?" asked a man in a smart suit. "Do they need to speak first?"

"Have ministers intervened on this area yet?" queried another.

The questioners introduced themselves as permanent representatives to the UN or members of their country's diplomatic mission to France. These were diplomats, entirely new to the climate negotiations. I wondered how they would find the COP. Would their diplomatic experience adequately prepare them for being thrown into the deep end of the climate negotiations' highest-profile session? It would be like playing the Super Bowl as your first pro football game. Even if you had the qualifications, the likelihood that you would do well was low.

When the discussion groups suspended for lunch, Janna and I hunted out UNESCO's cafeteria. Its fifth-floor location provided an excellent view of the Eiffel Tower standing brown against the pale sun of the winter landscape. Wandering around afterward, I couldn't get over the amusement of a photo booth on display called "mosaics of change." You stood against a green screen and pushed a button and a camera snapped a series of photos against culturally significant backgrounds: the fall of the Berlin Wall, press crews covering the Arab Spring, and so on. Janna and I tried to look appropriately awed, but the timing delay between the digital countdown and the actual flash rendered serious poses impossible. Mostly we looked a bit dim. It was difficult to grasp the gravity of the present historical moment, given the fact that we were laughing like idiots trying to put ourselves into others.

The moment did sink in later, as the LDC Group prepared for the preparatory meeting's closing speaker. Where the meeting would normally welcome the chairs of negotiating bodies—people like the cochairs Dan and Ahmed—the closing speaker for Paris was different. The headline guest was France's minister of foreign affairs, former prime minister and current COP president, Laurent Fabius. A distinguished-looking man in a navy suit complete with a pocket square, he fit the description of an "older French gentleman" if I knew nothing else about him—which I didn't. I had not had the pleasure of meeting former European leaders until today.

The hundreds of LDC negotiators reconvened for his appearance during our last session on Wednesday afternoon. From my seat next to Janna, I watched as Minister Fabius's security detail followed him down the center aisle. My gawking was interrupted by Giza signaling me from the top table. I shot him a confused expression, only to reluctantly follow the procession down the aisle to the front. Once there, sitting a seat away from the former French prime minister, I realized that Giza just wanted an entourage as large as that of the minister's.

Sure.

While the COP president welcomed us to Paris with grace, I looked out at the LDC Group and tried not to make eye contact with Janna as she snapped pictures on her iPhone. When flags raised for questions, I kept a speakers' list for Giza and generally tried to look intelligent. Minister Fabius gave his answers with gravitas, and the feeling of exam dread hit me again. This was it. The long-awaited COP would begin in four days, whether the world was ready or not.

And it was not.

———————

Marika arrived on Saturday, and that night we went out for some fine dining prior to the official start of what Janna had already dubbed "fat COP" given the ease of access to French pastries. I had always found French food as something to behold. I hadn't experienced a bad meal in the capital—gorgeous salads, macarons that melted in your mouth, bread, and a thousand cheeses. Marika and Janna had both lived in France, so whenever Janna Yelped vegetarian restaurants, I followed her toward what sounded like an excellent meal. Spending time with Marika and Janna felt more like hanging out with close friends than work colleagues now anyway, given the sheer amount of time we spent together.

We got off the metro at République and were immediately confronted: Candles and flowers. Rows and rows of messages to those so abruptly departed in the terrorist attacks of two weeks ago. In a coordinated series of suicide bombs, mass shootings, and hostage takings, terrorists had killed 130 people in France. The news broke just before Janna and I left for Paris. They were dining in cafés on a Saturday night—just like us, exactly one block from the vegetarian restaurant where we now sat.

I remembered where I was on 9/11, fourteen years old. Mom was driving our car pool down the hill to school. Even though a newscaster shouted over the car radio, Mom kept turning up the volume. "Be quiet! I think we're under attack," she yelled.

I didn't understand what she meant until we got to school and watched hour after hour of breaking news coverage. Our teachers abandoned their lesson plans, and students didn't bother to change classrooms at the sound of the bell. We all sat staring at the screens.

I remembered the smoke of the towers, buildings I had seen before during the bustle of a New York trip. I remembered the second plane, the people jumping from the skyscrapers hand in hand, and the sinking pit twisting my stomach in both fear and despair.

Now, I looked at the boarded-up café windows, the blast burns, the shoes left on the sidewalks and felt the same things, the shock and horror at how transient life can be, how quickly it can end and how painful the grief of those bereaved is. In preparation for COP 21, France had deployed the full sweep of its diplomacy, inviting every head of state and government to the talks. Given the terrorist attacks, many thought none would show over security concerns.

They were wrong.

Rather than negatively affecting the numbers, the attacks had bolstered them. Every leader would come to show solidarity. COP 21 was now predicted to trump even the 2014 New York Climate Summit as the single greatest gathering in history of heads of state concerning climate change.

The following morning, Marika, Janna, and I set out to get a pre-opening glimpse of the conference center. We took the metro northwest before going aboveground to catch a shuttle. Though I had been told the area of Le Bourget wasn't typically associated with security, the shuttle drove down carless roads, passed sidewalks swept clean. Side streets had been fenced off, with police directing motorists to turn around. The dry cleaners, supermarkets, and clothing stores along the route were either closed or nearly empty.

The French had certainly prepared for what was coming. Learning from the mistakes of the failed 2009 negotiations in Copenhagen, they had organized the Leaders Summit to headline the negotiations rather than end them. Over 150 heads of state and government were expected tomorrow.

When we stepped down from the shuttle, columns with the flags of all the world's nations greeted us. We quickly completed the small queues at registration. Our coats turned up against the chill, we wandered the near-vacant conference center, familiarizing ourselves with the venue, locating the nearly impossible to find, unmarked LDC office, and counting the cafés.

The place was an old airfield, with buildings both temporary and permanent laid out in a grid along either side of a runway that was now a pedestrian thoroughfare. Café chains had set up pop-up shops along the strip and promised all the pastries we could eat.

Inside the cavernous buildings were the typical windowless meeting rooms that were now so normal to me—the white tables shaped in a square or arranged classroom style, the tabletop microphones, the gray conference room chairs—all here to meet us once again. The corridors and open spaces inside the hangars were decorated with colorful animal cutouts. Janna and I took silly selfies with a red elk and a purple polar bear.

Having walked the length of the airstrip, we made it to the glistening miniature Eiffel Tower lit up in the dark. Upon closer inspection, I realized the tower was constructed of red metal deck chairs.

"Wow—this is so cool!" I shouted, waving Marika and Janna over. There was no one to hear us as we chatted our way back to the exit.

The difference between our Sunday evening ramble and the crush of Monday morning was startling. I had never seen security like this at a COP. With every head of state come to Paris, the crowds were astronomical. The plenaries required not only a PARTY badge but also additional clearance in the form of a red ACCESS GRANTED pass, limited to four per country.

The number seemed laughable given the sheer mass present. Google had the official count at thirty-eight thousand. This was more than triple the eleven thousand in Lima last year, which drew three thousand more than the negotiations in Warsaw had the year before. The corridors brimmed with people who looked utterly lost.

"Tourists," Bubu tsked after we said our good mornings outside the LDC office.

By now I could tell the difference between someone who had never been confronted with the bewildering agendas of the UN climate negotiations before and someone who had. Those who had were able to match their interest topic with the correct agenda item and locate the corresponding meeting room on the live screens. Those who hadn't, though, stared at the streaming schedule feed with a blank expression, hovered around the documents counter asking questions, and eventually gave up and sat looking at their phones in the cafeteria.

"Good luck," I said to Bubu as I pushed out into the crowd.

The additional security proved challenging for everyone. Just before 1:00 PM my phone buzzed.

"Hi, Marika," I answered. I clicked my computer closed and headed down the row of the room I was working in. "Sorry, lost track of time. I'm on my way to the LDC coordination meeting now."

"Don't bother," she said frantically. "We can't get in."

"What?!"

"The secretariat's assigned us a room in the restricted area, which none of us have security clearance for," she huffed. "Giza's been told he's not allowed. The other LDC coordinators aren't either. Tell everyone you see the meeting's been called off!"

Even with the Leaders Summit kicking off in the restricted area, the article-based spin-off groups began convening. Achala accompanied Pa Ousman and Bubu through the security perimeter and texted photos of President Obama sitting at the États-Unis table, which was within selfie distance of La Gambie's under the French alphabet.

Walking to the finance spin-off group, Janna reported being pushed out of the way by Secretary of State John Kerry's security detail. I, on the other side of the conference center, spotted former vice president Al Gore amid a swarm of media microphones. I made way with a shock of recognition. I knew him primarily as the face of *An Inconvenient Truth*—the film I rejoiced in so much that I spent spring break of my undergrad sophomore year putting on a screening back home in Kelso High School's auditorium.

After lunch, I consulted the live screens and navigated through the mob toward the technology development and transfer spin-off group. Once the door closed behind me, I nearly laughed. Given the crush outside, I expected a full house of newcomers to the technology negotiations. But there were only a few additions to the faces who had convened a month ago in Bonn. I shuffled around to sit next to Fred and Stella, whose presence I had grown quite comfortable in. Over the past several months, I had spent most of my negotiating time with them, rather than the chair, focused on best representing the LDC's interests in the technology portion of the draft treaty.

Without much ado, the discussion picked up where it had ended in October. This same group had spent the year on this one article. Our first meeting was that dreadful Geneva one in February, where the technology text swelled from the two pages of Lima to five. Five pages of treaty language about technology transfer were more than even I wanted to read. We reconvened in Bonn over June, August, and October, talking these down to just seven paragraphs—each of which I could now define with a single word. We were getting somewhere, though, and now that we were reassembled, I hoped we could finally agree.

As remained the style of the entire draft treaty, disagreement was reflected in a series of options. Each of our seven paragraphs had at least two—though the most used choice for Option 2 was "No text," which meant deleting the entire paragraph. Our aim was to narrow the options down to clean text—or at the very least to as few options as possible—before handing the article over to ministers the following week.

Discussions kicked off, and evening rolled around before long. The technology negotiations suspended for the 6:00 PM coordination slot, and I found Janna and Marika for dinner and gossip before heading to the 7:00 PM LDC coordination meeting. Now that the LDC office was labeled, Marika reported cramped conditions.

"There are just so many people—the Gambians came by, and there's the crew from IIED. Apparently Angola has its own office, only no one can find it, so they all came to ask for directions. The chair and Achala have come and gone. Have you seen Katharine?"

Marika asked this as we chose a line to stand in, the large CRÊPES sign ahead looking promising. Janna went off happily in the direction of stinky cheese. She kept telling me it was a French delicacy, but nothing I wanted to eat smelled like wet dog.

"No, not since lunch," I answered. Katharine was an IIED colleague who had come to Paris to help Giza manage his media commitments.

"Well, she wants to prep Giza for an interview later, if you see either of them. Oh, and Achala had some models stop by for her."

Achala had been recently featured in a *Vogue* piece about climate warriors.

I laughed. "Really?! Fashion models stopped by the LDC office?"

"I know. That's something I never thought I'd see." Marika smiled.

The cafeteria had a glass wall out to the pedestrian thoroughfare, so we people watched while we ate. I tried, and failed, to not make faces at Janna whenever I caught a whiff of her food choice. Delegates, students, businessmen, activists, diplomats, news crews, and personal assistants streamed by. Every now and again a group of cameras and fluffy gray microphones moving at top speed cut through the crowd. Sometimes I recognized the person who drew their attention. Most of the time I didn't.

Attendance at the 7:00 PM LDC coordination meeting, which at any other COP would have been low, boasted record numbers. Word had spread that coordination meetings were a good way of getting caught up on the events of the day and the expectations for tomorrow.

"Is finance meeting again tonight?" I asked Janna after Giza wrapped things up.

She stretched forward over the table in front of her and nodded. "We start at nine o'clock. What about technology?"

"Yup," I yawned. "It's the same for us."

Marika joined us from across the room.

"Late nights from day one, boss," I said, catching her up. In my four-COP history, this was the first where Monday ended in an evening session. "You'll have to head back without us."

"Good luck, guys. Text me when you get back to the hotel," Marika said when we parted ways—she to the COP shuttles and we to our respective negotiating rooms. I didn't see either of them again until the next morning, as technology outlasted finance that evening, and Fred and I barely caught the last metro back to the hotel.

This schedule continued for much of the first week. I never appreciated the ease of my London life until I worked six sixteen-hour days. The long hours were paying off, though. In the technology negotiations, only one paragraph had options remaining. The others still had brackets of disagreement, but at least we were all working from the same starting point. After a year of negotiating, I was glad things on technology were finally coming together.

On Friday morning, the cochairs issued an updated version of the draft treaty that consolidated the progress of the spin-off groups. The fifty-four pages we came to Paris with were down to thirty-eight. We read through the new version back in the room devoted to the 2015 agreement, whose furnishings had undergone a massive upgrade. An enormous square made of tables was bordered by three rows of cushy white chairs suitable for leadership. A wide aisle set apart three additional rows of regular seats. Screens hung overhead to display a close-up of whoever was speaking, provided by the cameramen stationed at the four corners of the room.

The read-through's aim was to identify missing elements and inconsistencies in the text. Delegates talked through the afternoon. I sat stunned in the audience, working through the articles as negotiators took the floor. Perhaps we were finally getting somewhere. As the hours stretched, the interventions grew shorter, and more and more people trickled out. By early evening, we had heard as much as possible.

"Thank you, delegates," the cochairs said after the last of the speakers finished. "We have come a long way. The work you have done allows us to pass the best-structured text possible on to the COP president."

Those left around the table clapped, their jobs partially done. The negotiations would move up a level for the political decision-making of week two.

"Dinner out this evening?" Marika asked when she found me after the discussion dispersed. A week had passed since I had set foot anywhere other than the conference center or our hotel. "Like out, out—in the actual city."

"Yes, please." I beamed at the thought.

We exited the metro across from Notre Dame, the cathedral lit up with white light and shimmering in the evening air. As it had on every other occasion I stood before it, the historic structure inspired the soundtrack of Disney's *The Hunchback of Notre Dame* to start running through my head. Unfortunately for Marika, I hadn't slept enough to control my impulse to sing along.

"Morning in Paris, the city awakes to the bells of Notre Dame!" I belted out. "The fisherman fishes, the baker man bakes, to the bells of Notre Dame!"

Embarrassed, Marika quickened her pace and acted like we had never met. A ridiculous idea. By now, our friendship had to be clearly recognizable. When I caught up to her, we sent a selfie to Tom before choosing a café. Despite my lack of French, I liked Paris. Though it wasn't the season, we walked around the islands of the Seine eating ice cream. Both of us knew this would be our only free evening for the foreseeable future. Upon Marika's request, I tried to keep the singing to a minimum.

The negotiators gathered in the red-roofed plenary on Saturday morning to watch the COP president officially take the reins of the 2015 agreement's negotiations. The secretariat carried heavy-looking stacks of paper down the aisles as Minister Fabius looked out over the crowd.

"Nothing has been decided until everything has been decided," Minister Fabius said through the interpreter—I listened in via the UN's standard-issue headphones. "We will be inclusive and transparent. No country, group, or issue will be left behind," he promised.

I remembered those last nights in Lima waiting outside the COP presidency's office hearing that the United States and China had gone in. It seemed the French were keen to learn from that mistake. A member of the secretariat was making her way down The Gambia's row. After she handed a small stack of paper to Pa Ousman and Bubu, she gave one to me as well.

It read *Draft Paris Agreement* at the top.

I blinked, trying to process.

What I had thought was impossible, the world had put on paper. The draft treaty even had a name—the Paris Agreement. The despair I had carried since Lima shifted slightly toward a lighter skepticism. Perhaps there was a tangible landing ground. Perhaps a result was possible. In my hands was a draft treaty that could end four years of negotiation, that plausibly, in less than a week, the UN could adopt into reality.

If the world wanted it to.

Minister Fabius's remarks ended, and flags went up. Reassured, the blocs praised the progress made so far with encouraging words of their own. "The world is watching," said a representative of the European Union. "We are here to adopt an agreement that is both applicable to all and acceptable to all."

On behalf of the G77 and China, South Africa ended its statement by echoing the words of Nelson Mandela, "It always seems impossible until it is done."

Text in hand, I followed Pa Ousman and Bubu out of the plenary. With all their side conversations with other delegates, I beat them to the LDC office and sat down next to Janna on the floor. We knew chairs would be in short supply once everyone assembled. I counted the familiar faces as they joined us, thinking back over how far we had come since Pa Ousman, Bubu, Achala, and I huddled in Durban's orange-carpeted LDC office. Now, Pa Ousman was an honorable minister. Ian was an ambassador. And we had the draft Paris Agreement to read through.

The LDC chair eventually called everyone in the room to order. Giza took us through the articles, pointing out where the group should focus its efforts and calling on the lead coordinator of each issue to pitch in. "In the preamble," he said, "we need to fight to keep the stand-alone paragraph recognizing the specific situations of the LDCs. This is the most important reference, but we should work to maintain the others made throughout. Honorable Minister, how does Purpose look to you?"

"For Article 2," Pa Ousman said, "we must retain holding temperature increase to 1.5 degrees. I've stressed this throughout the ministerials. For us this is nonnegotiable. It's a matter of survival." He ducked his head back down to continue reading through his glasses.

"Good. Sandra, take us through the mitigation articles."

"On mitigation, we need to push for collective efforts to be consistent with 1.5 degrees. We want a reference to emissions peaking or the numbers associated with that. Getting either will be a hard fight. The principle of no backsliding is more important. Each successive emissions reduction commitment must progress beyond the previous. This should be tied to a five-year review in the global stocktake—where's that?"

Ian answered her. "By the new numbering, it's Article 14."

Some people flipped ahead to the article and section devoted to reviewing collective action.

"Yes," Sandra went on. "This is our best hope of raising ambition over time."

"We can push no backsliding under Article 15 with the compliance mechanism as well," Achala came in.

"Wait, let's go through the articles in order," Giza interrupted. "Where's Thinley? Thinley, take us through Adaptation."

"The G77 has been strong on what's now Article 7," Thinley answered, beginning his review of the adaptation article. Pages turned and pens scribbled notes in the margins of the freshly printed text. The draft absorbed us, and the room quietly waited to hear the analysis of each lead negotiator. More people slipped into the room. They took seats on the floor or on top of the filing cabinets or leaned against any available wall space, listening.

"Maintaining clear references to implementing adaptation actions and support will be the big asks for next week's negotiations. Provisions for adaptation planning are currently in. It will be hard to keep both these and the references to the Adaptation Committee."

"Good," Giza turned to Ian. "Over to you, Ambassador, for Article 8."

"Well, let me start with the positives," Ian huffed. "We have loss and damage as its own article, which puts us on good footing for maintaining it. But we've had no substantial discussion of loss and damage this week, so I don't know where these streamlined paragraphs have come from. We'll need to work to get our proposals incorporated back in. We have solid grounds to argue this."

When Giza moved on to Article 9 and Evans, the lead finance coordinator, prepared to give his take, Janna leaned over to whisper the troubles of the finance negotiations to me. The references to scaling up climate finance from 2020's $100 billion per year floor were hanging by a thread, as were the mentions of special recognition for the LDCs.

After hearing Janna's assessment, I crouched forward to consult with Stella and Fred before the chair called them to take us through the technology article, which would be next. Fred, presenting what was now Article 10, explained that technology was down to six paragraphs, which retained most of the points the LDC Group found important.

If nothing else, the draft Paris Agreement met the mark I had focused on for the past year. I had noted the many issues of importance to the LDC Group as we went along. Ensuring that temperature rise stayed below 1.5 degrees Celsius, that money was provided as promised, that loss and damage was accounted for, and that the circumstances of the poorest were acknowledged were still options in the text. How could the treaty *not* have these?

"It will be over to the ministers and ambassadors now," Giza said, as if attuned to my wavelength on next steps. He looked to Ian and Pa Ousman, the read-through now completed.

The conversation moved on to assigning marching orders for the remainder of the negotiations. After everyone had left and we finished a cursory cleanup of the LDC office, Janna, Marika, and I left the venue together, drifting out with the steady flow of people leaving with what might be the future climate treaty in hand.

———

Tomorrow was Sunday, when the conference center would close for the mandated day of rest. I had big plans to do absolutely nothing: sleep

without setting an alarm and run along the canal. The first week of negotiations had proved a long one. I felt more drained than I'd anticipated.

The running part of the plan didn't happen until late. Long after
the sun set, my bulky, multilayered gear kept me insulated even when
jogging along the December water. My shoes squeaked against the wet
stone and pumped out a slow rhythm. My legs, turned jelly from a week
of sitting, rejoiced in the pull of strides. When they came without effort,
thoughts surfaced to fill the space.

The stunned relief I felt at holding the draft treaty had faded with
study. I started thinking through my growing list of the things the text
didn't have. Countries were committing only to communicate their
emissions reduction targets, not to achieve them. The question of
which countries would do what and who would pay weren't answered
yet. The limit set for global temperature rise remained a disturbing
choice with 1.5 degrees Celsius *or* 2 degrees Celsius still listed. So was
the principle that each successive commitment a country made had
to do more than its previous one. Countries couldn't backslide on
reducing emissions with changes in government if they were going
to solve the climate crisis. The article on compliance meant to ensure
this was so weak that I questioned whether it was worth the paper
necessary to print it.

My strides stretched in frustration and outpaced my lungs. I was
beyond the streetlights in the dark of an industrial lot, where the quiet
meant I could hear every irregular gasp I wheezed out. I slowed to a stop.

"I can't figure this out, Abba," I wheezed aloud. "What good is a
'climate deal' that will do next to nothing? It won't address the issue."

"Have you so little faith?"

Of course, I panted, thinking to spare my lungs. *Look at our track
record! We've been negotiating a "solution" to climate change my entire
life and all we've managed to improve is our understanding of the depth
of the trouble we're in. And they come, the least responsible, and the
rich run them down. And they agree—because what else can they do?
If they don't like it, they don't have the time or the resources to build
an opposition.*

"Couldn't you say the same for all 'lost causes'?"

*I suppose. But we're all going to die! From a long beating of unsurviv-
able disasters that kill the vulnerable first, while, what, the people most
responsible watch?!*

I had to work to slow my breathing, tears stinging at my ducts.

"My love, what will the agreement do?"

I didn't want to make sense of that. I walked, avoidant, seeing only
disconnected points. After several minutes they came together in coun-
terargument. *A treaty will strengthen international law that the world
has to work together,* I thought. *For the first time, ever, every country
will have to say how they intend to cut emissions.*

"And?"

*Hopefully, they'll have to do more as time passes. Cutting pollution
until we get to net zero—even if it's too late,* I muttered in angry aside.

He pressed unruffled, *"And?"*

*There will be some money for countries like the LDCs to deal with
adapting to the impact and the losses and damages they face. A world in
agreement is better than one where every country has to fend for itself.
Like any one nation could tackle our greatest threat alone.*

I looked up at the sky, searching for stars between the clouds. The
world needed a structure for combating the climate crisis and, despite
its flaws, the Paris Agreement could be it, the climate change treaty of
my generation. Our hope.

There was just the small matter of getting everyone to agree to it.

———

Monday's negotiations opened in the usual style with the high-level segment. Giza told the LDC Group to fully prepare its respective ministers—everything was at stake. I made a mental note to finish drafting Pa
Ousman's speech as I walked the pedestrian boulevard of the conference
center with Marika, Janna, and the rest of the morning crowd, grateful that
Honorable was counted as one of the LDC Group's most valuable assets
and would sit at the negotiating table today when the decision-making
kicked up to political leaders.

"Looks like they're convening a few of the thematic discussion
groups today," Janna said, scrolling through the live schedule on her

iPhone, "some ministerial consultations as well." I took out my own phone to follow the updates, both of us trailing Marika toward the LDC office. The ministerial consultations were titled with the thorniest issues: differentiating developed and developing countries' responsibilities, raising the level of ambition, and ramping up the provision of finance.

We found Pa Ousman, Giza, Bubu, and Achala huddled together after opening the office door. Seeing my inquisitive look, Bubu came over to fill me in. "Minister Fabius has asked Pa Ousman to facilitate the consultation on pre-2020 action."

"Oh, that's great!" I exclaimed. Because the Paris Agreement wouldn't come into effect until then, deciding what to do in the five-year interim was a critical piece of the puzzle.

"Is it?" Bubu surprised me. "If he's facilitating, it will take him away from negotiating. We'll lose him as a negotiator, and he's the most knowledgeable minister of the group."

I looked over at Pa Ousman sitting with Giza and Achala, thinking. The consultations would strain his already stretched time, but they would give him the chance to steer drafting the treaty's final text. I wondered which Pa Ousman would choose, which would be the most impactful for representing the LDCs.

"He'll do it," Bubu sighed, answering my next unspoken question. Being one of the fourteen ministers chosen to help Minister Fabius shape the agreement's final draft was a great opportunity to influence. It just wasn't the one Pa Ousman had planned for.

The ministerial consultations began that afternoon. Alongside the United Kingdom's secretary of state for energy and climate change, Pa Ousman facilitated the discussion of pre-2020 action. During a break, he delivered his statement at the high-level segment. I watched him on-screen in an overflow plenary, the hall itself too full and too restricted for me to enter.

"As of today, 185 countries have submitted contributions to the 2015 agreement," Pa Ousman spoke from the podium. "Though these contributions cover most emissions, the reductions they indicate are not enough."

The Gambia contributed 0.01 percent to global emissions, yet the country would commit to drastically reducing greenhouse gases from key sectors—44 percent by 2025—primarily through cutting methane associated with rice production.

"If my country can take these bold steps, then developed countries and those increasingly contributing to global emissions have the moral obligation to do more. We must ask ourselves, do we have the political will to act? Will we adopt an agreement here in Paris that facilitates the ratcheting up of our collective ambition?" Pa Ousman asked his peers.

I hoped they were listening.

The issues left on the table were questions only political leaders had the power to agree to. "Who does what?" and "Who pays for it?" were also the questions that had stymied negotiators for the past four years. The only way to come out of Paris with a new treaty was for them to finally agree to some answers. And we had just four days left.

The halls swelled with even more people as the media arrived en masse, gearing up for the COP's conclusion. There were cameras, lights, and sound engineers everywhere, and it wasn't long before colleagues back in London reported seeing glimpses of us on the BBC's coverage.

As the days progressed, the COP president implemented a stealthy strategy for keeping the numbers in the ministerial consultations down. With too many people to field productive negotiations during the day, the schedule began to creep. Each day, meeting start times pushed back from afternoon, to evening, to well into the night. Since I was a natural night owl, this suited me just fine. By Wednesday the group responsible for reading through the draft Paris Agreement wasn't called together until after 10:00 PM. Press conferences, coordination meetings, and side events continued as usual; the real negotiations happened by night.

The late schedule, and the additional security, really thinned the crowd.

We steadily switched to the graveyard shift. Giza called off the 1:00 PM coordination meetings, as everyone slept through them in making up from the night before. In the wee hours of Thursday morning, Minister Fabius told the thinly populated room that he would issue an updated text, which reflected the compromises made thus far, by that afternoon.

"Ask Marika to email the group," I heard Giza tell Achala as we shuffled out at 3:30 AM. "We'll convene at 7:00 PM and stay on through the night."

I had learned from previous experience that it would take the COP president many hours to come up with his next move. During that time, all that awaited me back at the conference center was just that—waiting. I slept most of the day in preparation and was geared up for the read-through when I arrived back at the conference center with Janna Thursday evening. Security issued another badge system to further restrict access, which Achala was well ahead of. She had additional LDC negotiators, Janna, me, and a handful of our IIED colleagues in before they were on to her.

Midnight broke and Friday began with the COP president, Minister Fabius, taking his seat. Around the table were prime ministers, including Tuvalu's, whom I recognized from his speech at the New York Climate Summit, and foreign ministers like US Secretary of State Kerry. I snapped his picture and sent texts off to Mom, Michelle, Erina, and Noelle.

Janna and I jittered with excitement next to each other.

Despite all my misgivings and disbelief, I couldn't help the surge of optimism. Unlike the past year, countries were now narrowing down the draft treaty's options by compromising and making trade-offs. The high-level negotiations were *actually* making progress toward agreement. The draft treaty was down to twenty-nine pages, and it seemed that countries—even though they had problems—genuinely wanted this. The US secretary of state was *negotiating*, for Pete's sake. I'd never seen such a highly ranked American diplomat sit around the table, much less at 3:00 AM.

With each passing hour, fewer delegations remained. Only the nations dedicated enough to sit around this table with the rest come 4:00 AM would have a say. I didn't know if this final push would be enough, but there was no way I was going to miss it, not having come this far. If this group got through the text, the world would have an agreement on climate change. The next night continued in the same style. When we hit an issue that seemed impassable, Minister Fabius sent the chief disputants into the hall to sort something out.

"Come back to us with agreed language, please," he facilitated. Consensus was brought back to the table again and again until no one spoke out against it.

We left the venue early Friday morning, taking taxis through the sleeping city back to our hotel, where I woke in the late afternoon. I checked my messages, emails, and the UNFCCC website for updated text that reflected the night's progress. My goal was to be back in the conference center before it came out. Janna, Marika, and I distributed printed copies in the LDC office as Giza conducted a read-through with the lead negotiators, and the cycle would repeat.

After three iterations of this, we met during the afternoon of Saturday, December 12, with heightened expectations. We had been through the twenty-nine pages of agreement text repeatedly, and we were out of time. As with all COPs, the negotiations were scheduled to close on Friday. Bets as to when we would actually finish were all over the place. I said Sunday. Bubu had said last night. Janna and Marika thought sometime today but guessed different hours.

When I arrived at the LDC office, Marika was distributing printed copies of the newest text. I took my usual position next to Janna on the floor, scanning as I sat. Over the last graveyard nights, negotiations had progressed on the issues the LDC Group deemed most important. I hoped this draft was better still. My glances at the relevant paragraphs noted forward movement, but I doubted the overall package was good enough. Giza's copy hit the table across from Pa Ousman.

"This works for us? That's it?" he asked.

I stared up at him dumbfounded—surprised more than anything else.

Pa Ousman locked eyes with Giza over his glasses. He nodded slowly, a small smile stretching his lips. Ian looked up as well.

"I think it's as good as we'll get," Giza said standing up. "The French have managed to work out something no one should object to. I see it working."

What?

I didn't believe it. I was holding a workable treaty? One that the LDCs were willing to agree to? I studied Pa Ousman's face carefully. He meant it.

Giza was gathering his things. "Come, let's see what the G77 has to say."

Janna and I walked behind the group of LDC negotiators to the G77 coordination meeting, which had taken over the room we'd spent our nights in. South Africa was speaking when we came in. China, India, Brazil, and the others sat around the table. They were having the same conversation the LDC Group had just finished. Pa Ousman and Giza claimed seats at an open corner to voice support for the text as it stood.

At this point, none of what was happening fit any of my past experiences. I had no frame to determine if we were close to final agreement or not. In the midnight negotiations, the diplomatic speak was so neutral that I could logically draw either conclusion. The jitteriness I felt during that first all-nighter resurged, with a mix of both fear of failure and excitement at success.

Yet, here they were nodding too.

China was saying the agreement worked for them. South Africa was congratulating the room on all their hard work, the effort it had taken to reach this conclusion.

This agreement.

I grabbed Janna's arm as she grabbed mine. Was this it? Had the UN done it?

And then the clapping started. The G77 would agree.

"Do you think the developed countries will go for it?" Janna asked as we stood surrounded by people shaking hands.

"I can't imagine they would oppose it. Not if the G77 agrees. But who knows at this point?" I marveled. Following the flow of delegates out of the meeting, we ran into Marika in the corridor. She had come looking for us.

"Do you hear the singing?" she asked. I strained my ears and heard a low chorus of voices coming from down the hall.

"That must be AOSIS," Janna grinned. "I guess the island states will go for it too."

We headed for the plenary to see what the rest of the world had to say.

———————

The buzzing of people in the corridors rose to a fever pitch when the live screens flashed the start time and that additional security passes were needed for the main plenary hall. I knew The Gambia's four seats were full. Janna was already on to Thinley about Bhutan's seats. I lost her in the crowd and was searching, without success, for an empty chair when my phone buzzed.

"It's only him and Dorji at Bhutan's table. Everyone else has left to make their flights," she thrilled in my ear. "Come on!"

I hung up and made my way to the front of the hall to find her.

Thinley smiled as I took the last back row seat of Bhutan's four designated chairs. "I have one more pass," he said.

"Really?!"

He nodded, handing it to me.

"Thank you!" I beamed, already dialing Marika. We had started this together four years ago. She needed to be here with us, should this be the end. I caught sight of Pa Ousman, Bubu, and Achala sitting at The Gambia's table as I made my way out. Pushing through the crowd and flashing my passes at security, I located Marika and we waded our way back into the hall together.

The plenary was jammed.

They had told us earlier that this old air hangar housed the plane that made the first transatlantic crossing. Its high ceiling was lit up in red and spanned an expanse large enough to cover what must have been two thousand people gathered inside. Every chair was taken. Marika followed me up to the front where Bhutan sat nearly first in the alphabetical order of countries. She, Janna, and I squeezed our three bums onto two seats as Minister Fabius took the top table.

"*Chers amis,*" he began, sitting down at his microphone. I slipped on headphones for the interpretation. The packed hall went quiet as everyone settled to attention.

"My dear friends," the interpreter said. "My apologies for having made you wait a little while. There were a few matters that had to be dealt with. The text you received this afternoon marks the culmination of a great deal of effort. I hope you've had time to discuss this amongst your respective groups . . ."

I leaned forward to give Marika some shoulder room next to me, smiling as I remembered the cheers in G77.

Minister Fabius handed the floor over to the legal and linguistic review team. I fidgeted as they talked through the lawyers' task, which involved synchronizing acronyms and italicization between the different-language versions of the draft agreement. The deputy executive secretary was up next. He was a small man whom we rarely saw in the daily negotiations, though he often sat at the top table during the closing plenaries.

He spoke in English to the crowd.

"Thank you, Mr. President. As a result of the finalization of documents in haste by colleagues who had not slept for days, a number of errors regrettably were not detected in the document as it was being finalized in the early hours of this morning. The secretariat regrets the errors, and I apologize for the oversights in moving from one document to the other. I will now read a list of technical corrections to the text . . ."

I reached for a pen to mark the corrections, opening the printed copy I'd held since walking out of the LDC office this morning. I had signed my name across the top line, a habit learned from having to relinquish my copy to higher-ups too many times. Under my name, the decision read **ADOPTION OF THE PARIS AGREEMENT** in bold letters.

"In preambular paragraph 10 after 1.5 degrees Celsius should be added 'above pre-industrial levels'; in paragraph 31a on page 5, 'common methodologies and metrics' should instead be 'methodologies and common metrics'; duplication: paragraph 35 and 37 there is a duplication that should be removed; in paragraph 54 . . ."

The corrections were coming too fast for my brain to process and my fingers to flip. After the first few, I stopped trying and instead locked eyes with the words **PARIS AGREEMENT**, drinking them in. The treaty was far from perfect. Yet knowing the LDCs would accept the deal, all I wanted was for the rest of the world to agree. In and of itself, the Paris Agreement would not solve the climate crisis—but without it, we didn't have a prayer. After everything we had gone through, I wanted the UN to adopt the historic treaty. I wanted this to be it, the final moments of a long journey.

"*Chers amis*," Minister Fabius said, bringing me back to the present. "I assure you these are technical rather than material corrections. I would like to transfer the draft agreement to the COP for adoption. We will adjourn this committee and immediately resume the COP," he said, procedurally transferring from the committee to COP, both of which he led.

Nobody moved. It felt like no one breathed. I looked over at Marika and Janna just to make sure they were still there. They were leaning forward in their seats, just like I was.

Minister Fabius read out the opening of the COP.

"I will now invite the COP to proceed, and I will give the floor to all those who wish to take it. I want that to be absolutely clear. I now invite the COP to adopt the Paris Agreement, which is in the decision." He paused.

"And I look out to the room, and I see the reaction is positive . . ." Four thousand eyes blinked at him.

"I see no objections. The Paris Agreement is adopted." He raised and lowered his gavel for a single shocked beat of silence.

After all that waiting, the final words happened so fast.

We rose to our feet in the bedlam, tears streaming from many eyes. Minister Fabius stood and put his hand over his heart in thanks before joining hands with those at the top table, UNFCCC executive secretary Christiana Figueres to his right. They raised their interlinked arms as we cheered and clapped with hands held high.

The screens that hung overhead panned away from the top table and circled the room. Al Gore in the front row was first on camera, then across to the South African delegation, Chinese representatives on their feet, John Kerry through the crowd. Back at the top table, Christiana gave two big thumbs up before pointing to the room. "It's you," she mouthed to the unrelenting celebration.

I hugged Marika and Janna tight and couldn't help the laughter that poured out loud.

We had done it!

For the first time in history, all nations had agreed to act on climate change in a legally binding treaty—under which they must bring down emissions, hold temperature rise to below 1.5 degrees, provide money as promised, and recognize the needs of the world's poorest countries. The UN had come through the standstills and the stalemates, the huddles and the last-minute, Hail Mary fixes. I had doubted the outcome and cursed the process, the sleepless nights, countless hours, and tears and sweat of the last four years. Yet we had structured the future international effort to combat climate change.

We had adopted the Paris Agreement.

And as I smiled at Marika and Janna, I was reminded again of why that mattered so much, was reminded of all the people I loved and why this moment was critical to protecting them.

But of course, that was not the end of our story.

16

INTENTIONS

London, United Kingdom
January 2016

I COULDN'T REMEMBER FEELING MORE UNCOMFORTABLE. In other circumstances, the contrast between my external and internal state would have amused me. The room I sat in was pleasantly warm, the chair I sat on cushy and comfortable. The house, peacefully still. Yet what I most wanted to do was cycle home through the sleeting downpour. It had taken nearly a year after my father's death, but tonight was the night: I was due to start therapy. Before this evening, my only counselors had been academic. I had underestimated how apprehensive this experience would make me feel.

I needed something, though. I still broke down crying in public for no attributable reason. A deep sadness that peaked in fits of tears had me wailing into pillows instead of sleeping, curled into a ball during the small hours of the morning. It happened less frequently now, but that only meant a week could pass without them. And I was slimmer than ever. Depression made a great appetite suppressant. Like everything else, food just seemed so pointless.

Marika had noticed. "How are you doing?"

Even post-Christmas holidays, I looked too mournful to pass for well. "Not great."

"Maybe you should talk to a professional? We have that counseling service on retainer," she reminded me. "You should call them."

I grimaced. I didn't want to talk to anyone—that was the problem. Though I loved Marika and Janna, I told myself that my London friendships were too new to cope with the magnitude of what was wrong. So, I had no one to talk to. But that was just an excuse, and I knew it. I spent the holiday months home with Noelle and hadn't said much then either. What was there to say? Dad was dead. Ours was a terrible relationship. I survived him, and like any storm weathered, his passing brought relief and a massive cleanup effort. I should just move on. Only, even I had to admit that the moving on part was proving much more difficult than anticipated. It was certainly taking a long time, and time in and of itself wasn't making pulling myself together any easier. Was it getting worse now that I didn't have the constant distraction of flying off to yet another round of negotiations?

I didn't know.

What I did know was that understanding that I needed help was one thing. Getting professional help felt like something else entirely, an unscalable effort, particularly when I couldn't even sleep at night.

"I'll give myself another month," I balked. "Maybe I'll feel better after February."

Marika stood both unpersuaded and persuasive. It took several conversations, but weeks rather than months later, I sat waiting in the warm, cushy room, next to a box of tissues, which I kept glaring at apprehensively.

This was going to suck.

My counselor was a gray-haired British woman who asked a lot of open-ended questions in a soft, authoritative tone. I usually found conversing with strangers enjoyable. I liked learning about people and was fond of asking open-ended questions myself. Only, as she constantly reminded me, I wasn't supposed to ask questions. I was supposed to answer them. Answer questions, from a stranger, about my very least favorite topic, the one I had worked my entire life to bury or conceal, for an entire hour.

It was like pulling teeth. I didn't know whom I felt sorrier for—her or me.

But it helped.

It helped because, after we talked through the whats and hows and delved into the whys, we got to the place I never could—the place past the pain, hopelessness, anger, and despair. The crippling grief of loss and the inability to imagine anything beyond. We got to plan what to do now, for a future sans Dad that acknowledged both the terror and unfulfillment yet could circumvent these pitfalls with a couple of recognitions.

First, it was inaccurate to think Dad was gone.

I'd unsuccessfully run from him my entire life. He always caught me, though—tripped me up. So much of my mental health was wrapped around him and his opinions. They had fueled a marathon of adolescence and young adulthood, and they remained inside my head, still there, still to be confronted for as long as he lived in my memory. Ironically there would be no running from him now. I would find forgiving him for all the pain he raked through my life an ongoing effort, regardless of his physical absence. No matter how vehemently denied, he remained, as did all the feelings he inspired. Repressing them did me no good. My subconscious was rather vocal, and if I wished to sleep through the night, allowing myself to feel the full spectrum of my emotions seemed a logical place to start.

This affirmation led to the second one, which materialized through the most painful parts of conversation. I didn't know what to remember. Of the hospital scenes, I oscillate between the punching and handholding. How could two such memories exist together? How could they both have been real? I've never known what to think about him. No one person was purely anything. People were multifaceted, as were our vast and multidimensional impacts on each other. While not pretending benefits or benevolence existed where they did not, I could, should I wish, allow myself to see memories usually overpowered by their stronger, more terrifying cousins.

My counselor and I combed through old remembrances looking for happy ones. It was an uncomfortable process, tiring work. But Dad had taught me how to ride a bike, to play racquetball. A long time ago, he used to do my hair. My love of African American culture began with Dad and the importance he placed on it, the extended family where it thrived. It lived in the books on his bookshelf and those I read as a child, the Black Barbie dolls he insisted we play with along with the white ones. The events he took us to, the protests and knowledge of both their value and permanence. The references and way of speaking. The music on the stereo. All indelible from my memories of him.

This was the easier, more amenable side. The things that hit uncomfortably close to home were harder to think about and halting to discuss: Dad's personal traits, so easily identifiable in myself. The fiercely guarded independence. The high tolerance for risk and craving for adventure. The strength and athleticism. The maddening drive to question. I liked those things about myself, and, without him, they probably wouldn't have manifested in the same way. Ultimately, finding some good was comforting, a welcome counternarrative to the memories I most prominently associated with Dad.

By far the easiest of the revelations, though, the one that actually got me talking and created inspiration, was our last discussion topic. "It's about identifying what helped you in your darkest moments and leaning into those in this one. What was it that kept you going?" my counselor asked.

That was easy.

I saw the love that held my sanity during the end of my father's life. Noelle picking me up from the Seattle ICU. My best friend, who was afraid of heights, determined to take me bouldering, insistent that I didn't spend the entirety of my twenty-eighth birthday in a hospital but doing something I enjoyed. Erina's family making multiple curbside deliveries to UW Medical Center. Bags of Whole Foods groceries, complete with my favorite honey-roasted peanut butter, care of the Craft family. Michelle, who, moments after Dad's death, skipped out of residency in Boston to board a cross-country flight, resolved to cook tuna melts for my grieving family and get me singing show tunes again. Marika's probing care and persuasion.

Without them, I wouldn't be here.

"There you go," my counselor said. "Those relationships are the ones you've built your life on, the ones that have seen you through. Lean into them. Appreciate them. Trust that, as they have been, your friends will be there for you now."

We talked through how I could go about telling them, how to be honest rather than silent. I was still the one to not write, text, or call back, who used distance and "business" as an excuse. The thought of being so free with my despair threw me. It was not a happy tale, the story of me and him. Why burden people with my grief? Why burden myself with the trial of telling? Perhaps it would get easier with practice. I certainly took her point that I did not want the same silence that haunted my family to inhibit my other relationships. The only way to gain help was to break it.

I remained entirely frustrated with my muteness. Since Dad had passed, the silence of my childhood had returned with a vengeance, just in time for the UN climate negotiations to hit mainstream news. After Paris, suddenly everyone wanted to talk. When I was asked about myself or even my work, getting genuine answers out made me so anxious that I shook. Not with the nervousness of public speaking but with other emotions, ones I didn't have names for. Each word uttered was hard fought. There wasn't even singing; simply repeating words memorized was too difficult.

I now envied those with the ability to speak freely, effortlessly filling journalists' answers to questions about the Paris Agreement with noise. How were they not scared? It didn't matter how much I wanted my people to honor their commitments and address the climate crisis; I knew another painful moment was coming as soon as *he* took office. It was only a matter of time.

In the end, it happened during a breezy London evening—afternoon in Washington, DC. Watching the news used to be a chore, a dull duty I felt bound to perform. Now, it was an emotional roller coaster featuring the world's most terrifying drama. Crises innumerable kept me retreating inside my head. President Donald Trump stood at a podium in the White House Rose Garden that brimmed with June sunlight. Then he started speaking, and for the first time in a long time, I transformed into the crazy flatmate who won't stop yelling at the TV, shocking even myself. I just couldn't help it.

"Thus, as of today," he said, "the United States will cease all implementation of the nonbinding Paris Accord and the draconian financial and economic burdens the agreement imposes on our country. This includes ending the implementation of the nationally determined contribution and, very importantly, the Green Climate Fund which is costing the United States a vast fortune."

What?!

No authoritarian international climate regime existed. No UN body had forced us to do anything. Every single thing Trump said about the Paris Agreement and the contribution the United States was supposed to make toward it was factually inaccurate. The US government had defined its own emissions reductions target, set its own pledge to the Green Climate Fund, and, in so doing, had weakened the treaty beyond contemporary use!

He kept talking about the cost to American jobs and the American economy as though there was no cost associated with doing nothing. As though no jobs or development were gained through addressing it. As though the climate crisis was not already costing Americans billions of dollars in damage every year, thousands of lives lost to more intense heat waves, droughts, and wildfires, more powerful hurricanes, and more crippling nor'easters.

I couldn't stand it!

"You're lying!" I kept shouting, the floodgates breaking open. The fear I had on Inauguration Day manifested into tearful embarrassment, a distress that finally broke through my lack of speech. At this rate, I would be hoarse by the time he finished.

"But the bottom line is that the Paris Accord is very unfair, at the highest level, to the United States," Trump said.

My mouth fell open. How utterly absurd.

Fortunately, Janna completely understood my sentiments. She took to social media in response to Trump's announcement where she posted pictures of Marika, herself, and me during Paris's closing plenary. In the shots, we're laughing—elated at the newly adopted treaty. Now she added, "F*** you Trump!" in thought bubbles over our heads.

I knew this was not the first time the United States had shaped an international climate agreement only to walk away from its responsibilities after its adoption. That didn't make dealing with the fact any easier. My country, the world's single largest contributor to climate change and foremost economy, would do nothing about the crisis it was most responsible for. The shame. The reckless disregard. We who had done the most to cause the problem and had the greatest means to fix it would not undertake the pledge we had written to bring down our emissions. Nor would we deliver the money we promised to the Green Climate Fund.

We would walk away.

We would be the only country on the planet not to act. We would let the poorest die from a problem of our making, while saying it wasn't fair—to us.

It brought it all rushing back. I remembered the shock and disbelief of learning about climate change in the first place, my horror that, in America, the problem was still not universally understood. I remembered the dread that followed as I continued to study, the sense that I should give in to fear, that it was all too late anyway. We had pushed the climate to unprecedented limits already, and the necessary amount of political will required to do anything about that didn't exist.

Based on these arguments, giving up seemed not only logical but smart.

But now, my mind also filled with all the reasons why we hadn't, why we couldn't, even now. There were answers and so many people worth protecting, the relationships I had built my life on. The friends I clung to with a wild and awed love.

I had a best friend. Noelle's and my friendship began when having a BFF was ubiquitous. In middle school, the term was used so frequently it could mean anything. I had lived long enough now to know that what we had was rare and precious. I had no secrets from Noelle. That didn't mean that I had told her everything. It meant that there was nothing I wouldn't tell her. I had spent my adult life hoping that I would end up on her sofa, just talking. I traveled the world with the assurance that I would always have a place in her house. This knowledge made me braver, able to take risks I wouldn't have taken otherwise.

And there were Michelle and Erina, friends I had laughed with across countries and continents. People who challenged me to see the world from perspectives I hadn't considered, taught me things and ways of living I hadn't known. I loved our shared sense of difference and the humor we found in it, that when I called Erina, "Shaniqua Lou" came up on her phone and that I still texted Michelle, "Morning, Hindi" and sent Erina the sushi Bitmoji. I loved the expanding horizons they brought to my life. They were such amazing people. Their fearless drive to achieve what they wanted inspired me to do the same. I saw them in the possibilities I imagined for myself.

Like Pa Ousman and Bubu, climate change impacted everyone I loved. It dictated the choices they could make and the futures they would have. Michelle wanted to practice medicine in New Orleans, a city already below sea level. Erina lived in New York. I checked in with her during Irene and Sandy, hurricanes that shut down the entire Eastern Seaboard. I wanted them to be safe, to not have to live with the fear that our changing climate would bring storms that were stronger and more frequent.

I wanted Noelle and Kevin to enjoy a Pacific Northwest that resembled what we had known as kids, to hike and camp in landscapes rich with animals that had survived there eons before we arrived. I wanted Kevin's contracting business to be successful and for him not to suffer the health impacts associated with working outside during the hottest years on record. I wanted Mom's garden to bear fruit in seasons that began and ended according to long-held patterns. I wanted another thirty years with her and another seventy years with Noelle, Michelle, and Erina in a world that we recognized.

My African American upbringing taught me that change did not come quickly or easily, that safety came at a cost paid in sacrifice and risk: The yearly marches. Granddad's stories of standing at the National Mall on that celebrated day in Washington, DC. Maynard, whom we saw every Thanksgiving, was there too, together with him in the crowd. The struggle was his too because the way things were hurt people he loved, and standing against it was the right thing to do. The people I cared about were worth protecting. Behind all the briefings and research, they were the reasons I marched and voted, invested in solutions rather than pollution.

I thought about Marika and Janna and facing the negotiations together. I thought about Becca, who was bringing her passion for sustainable energy to the utilities setor by working in public power. And I thought about what motivated us—all of us—and how its foundation was too simple and too inherent to overlook. It was perhaps the only thing powerful enough to rectify a climate in crisis.

17

REBELLION

London, United Kingdom
April 2019

WATERLOO BRIDGE, the four-lane viaduct that cut across the Thames with bus and bicycle lanes, was crowded. Just not in the usual way. Tourists were still taking pictures of the Houses of Parliament and the London Eye to the west and St Paul's Cathedral to the east. Cyclists still rode through their designated lines, but there were no buses, no cars, no traffic whatsoever. Instead, the thoroughfare brimmed with people.

Dismounting in the crowd, I inched my bike slowly forward, minding the children and the police standing arm to arm. Days ago, protesters had lined the roadways leading to the bridge, blocking vehicular traffic from entering. They set up tents, made a stage out of a truck they'd parked diagonally across the lane lines. A folk band was playing there now.

Rumor had it they were turned out over the government's inaction about the escalating climate crisis. That had me curious to see the protests occupying some of London's most iconic roads and bridges for myself. I wanted to know what exactly they stood for and what they hoped to achieve. I headed toward a white tent above which the rebellion's flag—an encircled X—waved. The word INDUCTIONS hung from its entrance. A woman was already speaking, so I joined the twenty or so others sitting cross-legged on the tarmac to listen.

"Extinction Rebellion, XR for short, is a nonviolent movement of direct action," the woman explained in a practiced way. I wondered how many times she had given this induction today.

"We have three demands. First, that the government must tell the truth and declare a climate and ecological emergency. Second, that the government must act now to halt biodiversity loss and reduce greenhouse gas emissions to net zero by 2025. And third, that the government must create and be led by the decisions of a citizens' assembly on climate and ecological justice."

These demands decorated the signs protesters held on either side of the bridge and drew me off my bike and into this tent. What a list! I wanted these things too, though I did have to look up what a citizens' assembly was. Google told me it meant a representative group of citizens who, after they were randomly selected from the population, learned about, deliberated on, and made recommendations in relation to a particular issue. XR wanted this practice to lead government decision-making about climate change, as they felt their elected officials weren't moving quickly enough. Citizens rather than politicians would decide to act faster in the face of this emergency. I listened attentively as she continued.

"We mean to achieve these demands by holding sites around London until our government acts. We break the law with nonviolent direct action. We know that nonviolent movements are twice as successful in achieving their aims and that their solutions are more long-lasting than those gained through violence."

I grinned. Thinking about nonviolent civil disobedience recalled strong memories of home.

"This is a safe space for everyone. In addition to being nonviolent, we do not consume drugs or alcohol. Families and children are welcome here, as are all of you," the woman smiled. "Do let me know if you have any questions. And do meet each other. We would love for all of you to be come as involved as you are able."

People around the circle applauded. I shook hands with the couple sitting next to me. When the circle broke and left the tent to make way for the next group of inductees, a volunteer stuck an XR sticker on my shoulder.

"Welcome to the rebellion," she said.

"Thank you." I beamed. I was thrilled to be living my values and a Star Wars fantasy at the same time.

It was a Thursday in mid-April. The protesters had held four iconic sites around London for the past four days. Though police were making arrests, Extinction Rebellion's numbers were growing. They intended to occupy until the government met their demands. And I would join them.

The following Wednesday, I lectured at the London International Development Centre. They wanted me to speak about where things stood with the UN climate negotiations. After the final gavel had gone down more than three years ago, the Paris Agreement rapidly entered into force with every country on Earth signing on to the historic treaty.

"The Paris Agreement is a landmark agreement in the international effort to combat climate change," I told the room packed with mostly students. "It was adopted in 2015, and it will come into effect next year, in 2020."

I paused to see if anyone had questions. From the second row, Marika smiled at me.

"Before I launch into the negotiations, I'll spend a moment on the state of our climate," I said, clicking to my next slide. The moving image projected behind me tracked global average temperatures from 1880 to the present, rising from blue to bright red across a world map. Next to it, the line of greenhouse gas emissions curved upward exponentially. The familiar graph painfully depicted the climate's respiratory crisis. It had spiked to new heights with the unprecedented atmospheric concentration of carbon dioxide.

"The world has warmed by one degree Celsius since the Industrial Revolution," I summarized, "and last year posted the highest greenhouse gas emissions on record with an increase of about 3 percent compared to the year before."

I clicked to the next slide.

"The five warmest years in recorded history were the last five years," I continued. "We see the impacts of the resulting climate change in record-breaking heat waves, reduced crop yields, the spread of infectious diseases, and increased extreme weather events, among many other things."

I stopped to scan the faces. I didn't want anyone passing out on me. These days, I rarely spotted a student taking this information as badly as I had thirteen years ago, when I sat where they were now.

"Under the Paris Agreement, every country defines its own nationally determined contribution to mitigate this climate change. These contributions detail how the country will reduce its greenhouse gas emissions beginning in 2020. Most of these commitments have a five-year time frame," I explained.

"Countries defined what they intended to put in their contributions in 2015. Since then, scientists have found that the cumulative pledges made under the Paris Agreement will not keep global warming below 1.5 degrees Celsius, beyond which most people's lives and livelihoods will be irreversibly impacted."

The graph behind me was shaped like a weather thermometer. An arrow at 1 degree Celsius read, "We are here." The mark at 1.5 degrees Celsius indicated the agreement's goal. Current pledges were on track to heat the atmosphere to 3 degrees Celsius—double the temperature rise we could withstand, three times what we were experiencing now.

Projections at 1.5 degrees Celsius showed a chance of year-round Arctic sea ice remaining and a chance at one in ten coral reefs surviving. At 2 degrees Celsius, Arctic sea ice and virtually all coral reefs, along with their fish—a quarter of which served to sustain countries like Tuvalu as their primary food source—entirely disappeared. A world warmed to 3 degrees Celsius was painful to imagine. In all probability, multimeter sea level rise would redraw borders the world over, and Tuvalu and other island nations would disappear.

"For the past three years, the UN climate negotiations have attempted to draft the rules that will bring the Paris Agreement to life," I continued. "Every five years, the UN will review global progress in limiting warming. After these reviews, countries will then need to make their next pledge to further reduce emissions. The hope is that governments base the ambition of these pledges on the findings of the review."

The principle that successive commitments under the Paris Agreement had to cut more emissions than their previous commitment stood. No backsliding had made it into the treaty after all. I moved to wrap things up.

"The UN, its Paris Agreement, and contributions to it are based on the decisions of national governments. These decisions are reliant on political will. If we are serious about getting governments to solve the climate crisis—to bringing emissions down far enough, fast enough— every single one of us will have to act."

I clicked to the next slide.

"I'm sure many of you recognize this protest?" My last slide held a picture of Extinction Rebellion, taken less than a mile from here.

Several people nodded.

"We're lucky enough to live in a place where the government is mandated to reflect the will of its people. Extinction Rebellion is calling for Parliament to bring emissions down to net zero by 2025. The eyes of the world are looking to citizen engagement." I'd already shown a picture of sixteen-year-old Greta Thunberg.

I smiled in conclusion. "I hope to see you on the streets!"

Marika waited for me after the lecture ended. We still worked together at IIED, and we still supported the LDC Group in the UN climate negotiations. The fifteen-minute walk back to the office took us past the squares and restaurants I'd spent the last five years cycling by every weekday, laughing with Marika and Janna in, as I made a home for myself in London.

"Thanks again for coming. It's always nice to have a friendly face in the crowd."

Marika smiled. "My pleasure. I enjoy your presentations. I really liked the questions you asked."

"Thanks," I laughed. "They're the questions I'm constantly asking myself about the negotiations." In truth, they were the questions that had haunted me since I was in Lima, the ones I still struggled with.

"'Do the compromises made between the world's 197 countries add up to an adequate response to climate change?'" I recited.

"No," we said together.

"*But*," I went on, "is an international system necessary to address an international problem? Does the UN give more influence to the most vulnerable than other forums? Do the systems set up by the Paris Agreement have the potential to get us where we need to go?"

"Yes, yes, and yes," I said, just as I had in the lecture.

Though their current inadequacy was maddening. Reckless. Unsafe. Without commitment from the United States. The Paris Agreement did not address the climate crisis. Not yet. Should politicians want to act, its systems had potential. That was about it. In the interim, the UN did continue to give more power to the world's poorest countries than the other places where decisions about climate change were made.

Marika sighed, nodding.

We were back at IIED. As a thank-you for lecturing, the organizers had offered me a small gift. After some thought, I'd requested a slice of vegan cake. Nothing said "Good job!" quite so well. Marika had laughed when they handed me the neatly boxed slice of chocolate. Now, she made tea and gathered cutlery. Obviously, we would share it. She hadn't attended just for the lecture. Besides, we had to fuel up for an evening of protesting.

After ten days of occupation, Extinction Rebellion was still going strong. We were nearly out of time, but there was hope for solving the climate crisis—even if it didn't lay foremost in the UN. There were solutions.

As I told the students, every single person would have to act. That was true. Yet the highest aim of these individual responses had to be shaping the collective one. The climate emergency was too far reaching and too advanced—only policy, both national and international, could bring emissions down far enough, fast enough. If everyone voted, protested, and divested like the lives of the people they loved depended on it, we could prevent more people from dying. Fortunately, there were ample opportunities to exercise these rights.

I joined the protests.

I wasn't the only one. When I arrived back at Waterloo Bridge, the crowd was chanting, "Show me what democracy looks like!"

"This is what democracy looks like!" I called in response.

Climate change was front-page news in the United Kingdom. It had been for days. Media mentions of the issue were at a five-year high. Children, inspired by Greta Thunberg in Sweden, walked out of school in protest on Fridays. Some of the strikers joined the Extinction Rebellion occupation as well, adding a belligerent exuberance that was fun to be around. Their signs were great.

FUCKING POLAR BEARS DYING!

I'VE SEEN SMARTER CABINETS AT IKEA.

CLIMATE CHANGE IS SCARIER THAN THE *CATS* MOVIE TRAILER!

The teens climbed on statues, hung off streetlights, screamed, and chanted. They were there in droves—thousands took to the streets of London. I hadn't felt an energy quite like theirs in a protest before. They moved between the sites still held by Extinction Rebellion, joined in the dancing that musicians inspired when they played from the truck-turned-stage still parked across the thoroughfare.

An even younger set was also present, minded by the concerned parents who had long been staples of my time at climate protests. I smiled at a dad who stood next to me.

"I love your sign," I said.

His four-year-old son sat on his shoulders, wearing a dinosaur bicycle helmet and a Superman cape. The sign his father carried featured a hand-drawn red T-rex and read, I LOVE DINOSAURS, NOT EXTINCTION.

"Thank you," he said. "It's William's sign."

"Great sign, William," I smiled up at the child. "I love dinosaurs too."

In response, William grinned shyly back at me.

With global temperatures up one degree Celsius, scientists now measured how rapidly humans were shifting the seasons. Species were going extinct at rates one thousand times those typical of Earth's past. Climate change was among the direct causes of biodiversity loss. Record nighttime heat now pushed people and animals beyond their tolerance, particularly in many of the world's most densely populated areas.

And we weren't cutting emissions fast enough. More would die.

Moving through the crowd, I passed a group of doctors and nurses wearing scrubs over their jackets. Their signs read, CLIMATE CHANGE IS A MEDICAL EMERGENCY. One shouted this as a police van arrived.

The protesters had formed an orderly queue of people sitting on the ground, waiting to be arrested. This was perhaps the most British thing I had ever seen. Brits loved waiting in a line. I didn't realize this persisted while they were breaking the law. Even the police respected the queue. That any practice could be so entrenched within a society blew me away.

Forcibly removing a protester from the middle of the road took approximately four policemen. The protesters didn't struggle. They just refused to move until the government acted. Given that the rebellion outnumbered the police force, arrests were happening slowly and with an amount of care and a lack of ammunition I was unaccustomed to. Legal aid volunteers and cameras watched every movement.

I saw an elderly man sit down in front of a younger one who was next up for arrest. I knew that jumping the queue was unacceptable behavior and moved forward to see what was going on. Seated, the older man turned to address the younger one—his words carried backward into the crowd.

"We've messed up your future enough, son," he said. "I'll go first."

The young man shook his hand.

I choked up. When the older man was eventually carried away, the crowd clapped and cheered around me.

A father in the arrest queue had laid a photo of his infant daughters next to him on the pavement. On a sign he had written, I'M JUST A FATHER WITH TWO CHILDREN WHO IS VERY FRIGHTENED FOR THEIR FUTURE. I wondered when he would be carried away.

In response to eleven days of occupation, Parliament declared a climate emergency the following week. Two months later, the United Kingdom became the first major economy to pass a law that would cut greenhouse gas emissions to net zero by 2050. This would not be quick enough, but it was the type of action we needed.

All of us.

EPILOGUE

THE CLIMATE crisis is impacting every person you love. Every. Single. One. We must all act if we are to stop it. The problem is too big for any one person, one government, or one country to solve alone. It will take all of us working together to shape our collective response. Climate change is the single greatest threat we have ever faced.

Vote. Any person without a plan to confront the climate emergency is not fit for leadership. Your elected officials must enact policies that cut greenhouse gas emissions to net zero, and they must work to bring communities together to stop climate change and address its unjust impacts.

Protest. By not acting with unrivaled urgency and determination, elected officials are endangering their citizens. This cannot go unmarked or unanswered. Our lives and the lives of our friends around the world are worth more than short-term profit. Make your government hear you.

Divest. Your time and your money should not fuel pollution. Nor should it subsidize fossil fuels. Wherever possible, finance the solutions that will sustain us—all of us. And let this choice register with those who have lost your investment. Start with the things that matter to you.

Let love guide your way. It is the only thing powerful enough to solve this crisis. Those you love are worth protecting.

Love is climate action.

———————

People are incredibly resilient, able to face and endure even the most devastating personal crises. Our ability to harm each other is rivaled only by our ability to heal each other. We can learn new ways of being and imagine different futures for ourselves. We can rediscover joy and belonging, find meaning and peace. Change is possible. In the United States, dialing or texting 988 will connect you to the Suicide and Crisis Lifeline, where trained counselors stand ready to listen and support. Resources for traumatized children and their families and communities are available at the National Child Traumatic Stress Network (www.nctsn.org). As the network reminds us, reporting abuse or neglect can protect a child and get help for a family. It may even save a child's life.

ACKNOWLEDGMENTS

LIFE IS A CULMINATION OF STORIES. I remain grateful to everyone who has contributed to mine.

Thank you to those who thought this story worth telling. To my editor, Alicia Sparrow, and her team at Chicago Review Press, your energy and enthusiasm took this tale from draft to vision. Thank you to the team at Fuse Literary, especially my literary agents: the indomitable Veronica Park, who was the first in the business to gamble on my potential and represent me with such enthusiasm, and Gordon Warnock, whose steadfast commitment saw me through.

My gratitude to the memoir's early readers, both those in the book business and those who are not: Maya Millett, whose generous and thoughtful notes paved the way. Damian McNicholl and Leo Barasi, who gifted me their time, energy, and opinions. Marika Weinhardt, Janna Tenzing, Erina Aoyama, Michelle Christopher, Chalayn Nagunst, Kay Craft, and Max Lampson—you believed in this story even when I did not. Without your support this memoir would not be.

Thank you to the International Institute for Environment and Development, which remains my research home. To the dedicated professors at the University of Washington and Brown University who educate students about the climate crisis. And to everyone protesting inaction.

My special thanks to the delegates of the Least Developed Countries Group, who work for climate justice every day.

And thank *you* for reading.

APPENDIX

THE LEAST DEVELOPED COUNTRIES

Afghanistan
Angola
Bangladesh
Benin
Bhutan
Burkina Faso
Burundi
Cambodia
Central African Republic
Chad
Comoros
Democratic Republic of Congo
Djibouti
Equatorial Guinea
Eritrea
Ethiopia

Gambia
Guinea
Guinea-Bissau
Haiti
Kiribati
Lao People's Democratic
    Republic
Lesotho
Liberia
Madagascar
Malawi
Mali
Mauritania
Mozambique
Myanmar
Nepal
Niger

Rwanda
São Tomé and Príncipe
Senegal
Sierra Leone
Solomon Islands
Somalia
South Sudan
Sudan
Tanzania
Timor-Leste
Togo
Tuvalu
Uganda
Vanuatu
Yemen
Zambia

The forty-eight Least Developed Countries, dated 2015. Current information about the LDC Group in the UN climate change negotiations is available here: https://www.ldc-climate.org/.

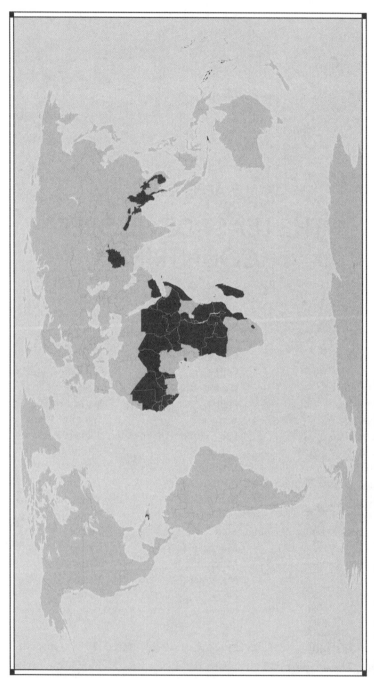

Map of the Least Developed Countries, highlighted.